The Aligned Enterprise

The Aligned Enterprise

A Practical Enterprise Architecture Methodology for Strategy-to-Execution Alignment

Jeff Davis

Kettera Press

Table of Contents

Enterprise Architecture Methodology

Chapter 1: Defining Enterprise Architecture 11
Chapter 2: Defining Success in Enterprise Architecture 21
Chapter 3: Methodology vs Framework 31
Chapter 4: Architecture Constraints (Three C's) 41
Chapter 5: Kettera Architecture Methodology 51

Business Architecture

Chapter 6: Organizational Goals Over Technology Goals 63
Chapter 7: Organization Strategy 73
Chapter 8: Marketing Domains 85
Chapter 9: Operations Domains 97
Chapter 10: Finance Domains 107
Chapter 11: Information Technology 117

Technical Architecture

Chapter 12: Application 129
Chapter 13: Data 139
Chapter 14: Integration 151
Chapter 15: Platform 161

Enterprise Architecture Artifacts

Chapter 16: Defining What Enterprise Architects Do 173

Chapter 17: Artifacts That Are Really Used 183
Chapter 18: Capability Maps 193
Chapter 19: Current vs Future State Diagrams 205
Chapter 20: Roadmaps (12 to 24 month) 215
Chapter 21: Application Portfolio 227
Chapter 22: Standards and Reference Frameworks 239
Chapter 23: Solution Architecture Diagrams 251
Chapter 24: Data Relationship Diagrams 263
Chapter 25: Technology Portfolio 275
Chapter 26: Architecture Decision Records 287

Enterprise Architecture Cadence

Chapter 27: Architecture Cadence - the Kinetic Difference 301
Chapter 28: Concrete Structure with Adaptability 311
Chapter 29: Innovation Pipeline 321
Chapter 30: Idea Generation 331
Chapter 31: Idea Refinement 341
Chapter 32: Value Analysis 351
Chapter 33: Prioritization 361
Chapter 34: Resource Planning 369
Chapter 35: Project Approval 379
Chapter 36: Transition 389

Conclusion

Chapter 37: Key Take-Aways 403
Chapter 38: What's Next 413

Enterprise Architecture Methodology

CHAPTER 1

Defining Enterprise Architecture

Enterprise Architecture is talked about everywhere but is one of the most misunderstood organizational disciplines, even by those who are practicing it. Some companies treat it as a list of technology rules. Others see it as a review board that signs off on projects. In many places, it's reduced to diagrams tucked away in a repository that are rarely opened and often forgotten. Each view holds a piece of the puzzle, but none of them tells the whole story of Enterprise Architecture (EA).

This chapter provides a practical, real-world definition of EA. It draws a clear line between EA and solution architecture, and between EA and day-to-day software development. The core idea that shapes this book is simple: Enterprise Architecture is a disciplined way to connect business strategy, day-to-day operations, technology choices, and long-term change. Using the Kettera EA Methodology as our anchor, we'll look at EA as a way of working, not a collection of documents stored on a server somewhere.

What Enterprise Architecture Is and Is Not

Enterprise Architecture is the disciplined practice of shaping how an organization is designed and how it changes over time. The "enterprise" here means more than IT systems. It covers business capabilities, everyday processes, data, applications, infrastructure, governance, and the links between all of them. Architecture, in this context, is about making deliberate choices. It asks practical questions: How should this organization work? Which capabilities does it need to strengthen or build? Where should it invest in systems? How do all these parts connect into something coherent?

Rather than focusing on a single department, EA looks across the entire organization. It ties long-term strategy to what teams actually do each day. When it works as intended, leaders can see the current state clearly, agree on a realistic future state, and map out how to move from one to the other. That map is not just a slide deck. It shapes budget decisions, sets priorities, and guides delivery work across departments.

What EA is not is just as important. It is not software development. Developers write code, set up systems, and deliver features. Their focus is on building and implementation. Architects may shape design patterns and standards, but they do not replace development teams. Likewise, EA is not the same as project management. Project managers track schedules, budgets, and risks for specific initiatives. Enterprise architects step back and ask whether those initiatives, taken together, are moving the organization in the right direction.

Many organizations confuse Enterprise Architecture with solution architecture. The distinction between the two is important. A solution architect designs a specific solution such as a new customer relationship management system or an integration between two applications. The scope involves less breadth than what the Enterprise Architect are responsible for. Enterprise architects look at how that solution fits into the broader application portfolio, data strategy, and business capability model. They consider not only whether the solution works, but whether it is the right choice given the organization's long-term plans.

A Working Definition

Here's what we mean in this book:

Enterprise Architecture is a clear, disciplined way to line up an organization's strategy, capabilities, processes, information, and technology so it delivers real, trackable results over time.

It's about making sure the big ideas connect to the daily work, and that the systems behind the scenes actually support the plan, not fight it. Strategy, operations, data and technology all have to pull in the same direction. Enterprise Architecture provides the structure to keep everything aligned and moving toward outcomes that are meaningful and measurable.

To begin with, EA is a methodology rather than just a framework or a set of templates. Frameworks offer structure and shared terminology, outlining which artifacts might be produced. A methodology goes further. It explains how the work actually gets done, including what steps are taken, who participates, how decisions are reached, and how outcomes are evaluated. The Kettera EA Methodology, introduced later in this section, concentrates on this hands-on, execution-focused aspect.

Alignment is the primary goal. Misalignment shows up in many ways: duplicate applications, inconsistent data definitions, overlapping processes, or technology spending that fails to support strategic goals. EA works to identify and correct these gaps. It establishes clear traceability from strategy to capabilities to system implementation. With that alignment in place, technology investments can be explained in business terms, and business initiatives rest on solid technical foundations.

Additionally, EA is outcome-driven. The goal is not to produce diagrams and flowcharts. It is to improve performance, reduce risk, control costs, and support growth. If architectural work cannot be tied to measurable business results, it turns into an academic exercise. The Kettera approach focuses on artifacts and a process that are actually used to make decisions, not documents created just to satisfy process rules.

The Scope of Enterprise Architecture

Enterprise Architecture reaches across the whole company, not just the IT department. Many people are surprised by that. They hear "architecture" and picture servers and software. This thinking is further reinforced by the fact that most Enterprise Architects have a technology background. Technology matters, of course, but it's only one piece of a much bigger picture.

Look at the business itself. EA dives into strategy, how the company makes money, who its customers are, how work actually gets done, and what financial limits shape decisions. It forces leaders to ask a hard question like "what are we truly better at than everyone else"? A retailer, for instance, might treat supply chain speed and precision as a secret weapon, while payroll is simply a routine task. Those two areas should not get the same time, money, or executive attention, and EA makes that clear.

Technology such as applications, data, integration approaches, and platforms all fall under the EA umbrella, as well. Architecture teams set guardrails for things like cloud configurations and security models. They also keep a running inventory of systems across the company, spotting duplicate tools and outdated technology before they drain resources. No system stands alone, even if teams like to think it does. Each one plugs into a broader environment, and changing one piece can ripple across the rest. That big-picture view is the real work of Enterprise Architecture.

Enterprise Architecture shows its real value when these perspectives come together. Imagine a financial services company preparing to launch a new digital product. A solution architect would focus on designing the required application components. An enterprise architect, by contrast, steps back to assess whether the current data architecture can handle real-time analytics, whether the integration layer will scale as demand grows, and whether regulatory reporting obligations are fully covered. They also consider how the new offering fits with the company's broader strategy and whether delivering it calls for new business capabilities.

Enterprise Architecture vs. Solution Architecture

Solution architecture focuses on a specific initiative. It turns requirements into a concrete technical design. The timeline usually matches a project lifecycle such as months, not years. Success means delivering a working solution that meets both functional and non-functional requirements.

Enterprise Architecture, by contrast, spans initiatives and stretches across longer timeframes. It sets guardrails and reference models that steer multiple projects. Instead of asking, "How should we build this system?" it asks, "What principles should guide all systems?" and "How does this project impact our overall portfolio?"

Picture a company rolling out five big tech initiatives at the same time. Each system might look solid on its own, carefully planned and built. Look closer, though, and issues appear. Without a shared enterprise view, teams may choose five separate ways to connect systems, copy customer data into mismatched formats, and quietly drive up support costs for years to come. The result is a patchwork environment that's hard to manage and even harder to fix. Enterprise Architecture steps in to prevent that mess. It aligns scattered projects and brings them under one clear direction, cutting down confusion and waste.

This only works if the roles work together. Enterprise architects set the bigger picture: guardrails, shared standards, and a long-term path forward. They focus on where the organization is headed strategically. Solution architects stay inside that frame while making the hands-on design calls for specific projects. When everyone knows who owns which decisions, tension is reduced and responsibility becomes obvious.

Architecture and Development: Drawing the Line

Confusion often arises around where architecture ends and development begins. In smaller organizations, architects are frequently drawn into coding work. At the same time, developers may end up making structural decisions simply because no formal architecture role exists. This works fine but even if the roles are filled by the same people, they should know which domain they are operating in for a given task and treat that work appropriately.

Architecture sets the structure and intent; development carries it out. For example, an enterprise architect might establish a rule that all customer data must live in a single master data management platform. The development team then builds or configures applications to align with that rule. When exceptions are needed, they are reviewed consciously instead of slipping in unnoticed.

None of this means the two groups work separately. Architects stay in regular contact with development teams and keep up with technical limits and new tools. Still, their focus stays on the system as a whole rather than on individual pieces of code. When they get absorbed in implementation details, they surrender the broader view the organization depends on.

Enterprise Architecture as a Continuous Practice

Enterprise Architecture is not a one-time transformation effort. Organizations change. Strategies shift. Mergers happen.

Technologies evolve. Architecture has to evolve with them.

The Kettera EA Methodology treats architecture as an ongoing rhythm rather than a one-off project. It includes regular review cycles, portfolio check-ins, and roadmap updates. Instead of producing a five-year blueprint that quickly becomes outdated, it focuses on rolling 12- to 24-month roadmaps that reflect current priorities and constraints. This rhythm ties directly to governance. Architectural decisions affect funding approvals, project prioritization, and standards enforcement. Without being built into governance processes, EA becomes advisory and easy to ignore. When integrated, it shapes real outcomes such as what gets built, what gets retired, and where resources go.

Practical Applications of Enterprise Architecture

Picture a hospital network squeezed by climbing operating expenses and patient information scattered across disconnected systems. An Enterprise Architecture team would start by laying out the organization's core capabilities such as patient intake, clinical care, billing, and reporting, and then pinpointing where things break down. They would also review the full stack of applications in use, calling out duplicate tools running in different clinics and the confusion that comes with them.

Next comes a clear picture of what "better" looks like. The architects sketch a future where patient records live in one shared platform, integrations follow consistent patterns, and reports line up cleanly with regulatory demands. They build a step-by-step plan to get there, weighing cost, risk, and day-to-day disruption and they use that plan to judge every proposed project. Does this move us closer to the target state or not?

Lower software licensing costs and cleaner, more reliable data show up gradually as systems are consolidated and standards are enforced. New clinics can plug into the environment faster because the groundwork is already in place, and fewer one-off fixes are needed. None of this happens just because someone drew a cool diagram. Real gains come from steady alignment and follow-through, year after year.

Key Takeaways

Enterprise Architecture is a structured discipline that aligns business strategy, capabilities, and technology to deliver measurable results. It looks across the entire enterprise rather than focusing on a single initiative. Unlike solution architecture, which centers on specific implementations, EA works at a broader scope and over a longer horizon. And unlike software development, its responsibility isn't building systems but guiding direction and coherence.

EA isn't defined by how many diagrams it produces or by strict loyalty to any single framework. What matters is impact: better alignment, less duplication, smarter investment choices, and clearer roadmaps. Methodologies such as the Kettera EA Methodology puts practice ahead of theory and steady rhythm ahead of static documentation.

CHAPTER 2

Defining Success in Enterprise Architecture

Enterprise Architecture usually gets explained through its artifacts like models, standards and multi-year roadmaps. Those things matter but they are not the main point. A company can fill shared drives with polished diagrams and still wrestle with duplicate platforms, stalled projects, and annoyed executives. Real success in EA shows up in results, not in how many documents the team publishes.

This chapter spells out what "success" actually looks like in Enterprise Architecture. Documentation and meetings are easy to mistake for value but activity alone changes nothing. Using the Kettera EA Methodology as a reference, we'll describe success in concrete, observable ways: smarter decisions, clearer investment bets, smoother delivery, and better alignment across teams. You'll see how EA proves its worth not in theory, but in how leaders choose, fund, and execute the work that matters.

Success Is Business-Centered, Not Architecture-Centered

A common misunderstanding about EA success is the idea that it revolves around the architecture and technology itself. When architects judge their work by how complete a framework looks, how polished a capability map appears, or how closely they follow a standard, they can lose sight of why the role exists at all. Enterprise Architecture is successful only when the organization performs better because architecture is in place.

That improvement can show up in several ways. Strategic initiatives move faster because dependencies were spotted early. Technology investments line up with business priorities instead of departmental preferences. Redundant applications are eliminated, cutting cost and complexity. Leadership gains clearer visibility into how current capabilities support future goals. These are business outcomes, not architectural ones.

Consider a retail organization planning to expand into digital channels. If Enterprise Architecture produces a 200-page document describing the current technology stack but fails to help the company spot integration gaps or required platform upgrades, the effort has failed, no matter how thorough the documentation. Documentation alone isn't value. If, however, EA clarifies which systems must be modernized, estimates the required investment, and lays out the work over 18 months to support the expansion, that is success, even if it was accomplished with 2 pages of documentation.

A modern methodology such as the Kettera EA Methodology

starts with organizational goals for this reason. Architecture always serves strategy.

Alignment as a Measurable Outcome

Alignment gets talked about constantly but almost no one pins it down to something meaningful. In Enterprise Architecture, it means one simple thing: business strategy and technology delivery move in the same direction, tied together by a visible thread. Ask leaders to spell out how a specific project advances stated goals. Can they do it? If the explanation falls apart, alignment isn't there.

You can spot alignment in the wild. Project portfolios mirror strategic priorities instead of last year's spending habits. Tech standards push the business forward. For example, backing scalable platforms when growth is the strategic goal. Gaps in critical capabilities surface early, before big initiatives absorb time and money. Budget conversations start to reflect architectural logic rather than politics.

The Kettera Methodology insists on traceability from strategy to capability to initiative. Every project has a reason to exist. Architects can name the capability being strengthened and the goal it serves. Executives might skip the architecture jargon, but they sense the fit. Decisions speed up because the yardstick is obvious and shared.

Alignment isn't abstract. You can see it play out in meeting agendas, funding decisions, and roadmap conversations. It shows up in what gets prioritized, what gets approved, and what drops off the list.

Decision Support as a Primary Indicator

Enterprise Architecture succeeds when it improves the quality and speed of decision-making. Those are the most practical ways to measure its value.

Organizations make hundreds of technology-related decisions each year. Which vendor should we choose? Should we build or buy? Can we retire this legacy system? Do we integrate or replace? Without architectural guidance, these decisions happen in isolation and short-term convenience often wins.

Effective EA runs on structure. People make choices inside a clear frame of reference instead of winging it. Agreed-upon standards shape which technologies get chosen. Capability maps show what's in scope and what isn't. Diagrams of today's systems and tomorrow's plans reveal ripple effects that no one intended. Architects don't block progress. They add perspective and offer context so leaders understand the trade-offs before they commit.

Imagine that the finance team wants a new analytics tool. In a company without solid architecture practices, approval might take minutes. Another contract gets signed. Another isolated data pool appears and no one asks how it fits with the rest of the system. Now imagine the same request in a company where architecture actually works. The proposal gets checked against the existing data model and integration rules, along with the platforms already in place. Does it connect cleanly? Does it duplicate something the company already owns? The answer might still be yes, buy it. But this time there are clear requirements for how it plugs into other systems and how it supports company-wide reporting standards. The point isn't to add bureaucracy but to make decisions with the right information.

A useful way to judge whether EA is working is to watch what happens before major investments are approved. Do leaders ask for architectural input early on, or only after plans are already set? When architecture is regularly brought into strategic discussions, before the big investments are in place, it shows that leaders value its perspective and trust its judgment.

Reduction of Complexity

Large organizations build up complexity over time. Applications overlap. Data definitions drift apart and integration points grow. Each extra layer adds cost and risk. Enterprise Architecture works when it cuts unnecessary complexity.

This doesn't require ripping out every old system or enforcing uniformity at any price. Big companies are complicated by nature. Some of that comes with the territory. The aim is not zero complexity; it's deliberate complexity that includes tools and processes that exist for clear, defensible reasons.

Modern methodologies such as as Kettera tackle sprawl with practical, structured tools like application portfolios and technology standards. When every application is listed and tied directly to a business capability, overlap in the systems is uncovered. The discussion changes. Instead of vague frustration leaders can point to evidence: three platforms handle the same task, and two of them can be phased out within 24 months. Facts replace opinion and decisions get more informed.

Results show up over time and redundant applications start to disappear. Data ownership becomes obvious and integration patterns get standardized across teams. Support costs drop. New hires ramp up faster because workflows make sense. Projects move forward with fewer redirections and less confusion.

Complexity never vanishes. What changes is control. The kind of complexity no one owns, the accidental kind, gradually shrinks, replaced by systems that exist on purpose and earn their keep.

Enabling Change, Not Slowing It Down

Enterprise Architecture often gets blamed for slowing innovation. Lengthy review cycles, inflexible standards, and theoretical debates can make it feel that way. When architecture becomes associated with red tape, people simply work around it and the value of EA declines. This can be true when implemented poorly but in reality, EA should do the opposite. Its job is to speed up change by bringing clarity and cutting down on avoidable rework.

If a future-state architecture is already mapped out, project teams don't have to invent the structure as they go. Clear standards mean procurement moves faster. Well-documented integration patterns allow developers to reuse what's already been proven instead of experimenting under deadline pressure.

Consider a healthcare organization rolling out a new patient engagement platform. Without architectural foresight, integration problems might not appear until late in the project, triggering delays and budget strain. With defined integration models and data standards in place, the team connects to established patterns and keeps moving because key decisions were settled earlier at the enterprise level.

In the Kettera approach, architecture works in a rhythm aligned with business planning cycles. This timing ensures that architecture changes alongside strategy. It is not a fixed set of rules. It is a living guide that helps ongoing change. If project teams describe architecture as helpful rather than a roadblock, that is a strong sign of success.

Governance That Adds Value

People often mistake governance for control. In the world of Enterprise Architecture, it shouldn't be about policing behavior; it should be about making things clear so teams can move with confidence. When governance works, people know where the guardrails are and why they exist.

Standards should be easy to find, easy to understand, and used the same way across teams. When something doesn't fit the standard, the exception process needs to be deliberate and visible, not handled through back-channels. Governance groups exist to help teams weigh trade-offs with open eyes and not to rubber-stamp rules without consideration. What's the point of governance if it shuts down judgment?

A strong architecture review does far more than stamp "approved" or "rejected" on a slide deck. It calls out real risks. It surfaces hidden dependencies. It makes sure decisions are written down. It clarifies who owns what. Architecture Decision Records capture the reasoning behind a choice such as the context, constraints, and alternatives that shaped it. Months later, when priorities shift or new information appears, teams can revisit that record and understand exactly how they got there instead of guessing at half-remembered conversations.

When governance works well, it feels steady and predictable. Project teams know what's expected and plan for it, so fewer issues catch them off guard. Tension between architecture and delivery teams eases as roles and standards stay clear, and that clarity builds trust. When governance is weak, friction shows up; when it's sound, confidence grows.

Measuring Success: Leading and Lagging Indicators

Defining success conceptually is useful, but organizations also need clear measures. Both leading and lagging indicators matter.

Lagging indicators include reductions in application count, lower infrastructure costs, fewer integration failures, and improved project delivery timelines. These metrics show long-term impact. They matter but they change slowly.

Leading indicators, on the other hand, reveal whether the architecture function is gaining momentum. Are architectural documents used in executive discussions? Are capability maps updated as strategy evolves? Do project charters reference future-state architectures? Is there a clear 12–24 month roadmap aligned with strategic goals?

The Kettera EA approach asks you to measure each kind of indicator so you can see the full picture. Big results often move slowly, and without smaller signs of progress you might think nothing's working. At the same time, a stack of reports and neatly filed documents doesn't mean you've changed anything in the real world. Looking at both sides provides a balanced view of the results.

Success isn't one number on a dashboard. It shows up as a trail of proof across different measures and milestones. One data point can mislead; a cluster of signals tells the real story. What's more convincing: a single spike, or steady movement backed by clear results? Real progress leaves patterns behind.

Cultural Integration

Enterprise Architecture does not exist on its own. It works alongside finance, operations, marketing, and technology. When EA is seen as an outside force trying to impose order, people have a tendency to push back. Success depends in part on how well architecture fits into the company's culture. Architects need to speak the language of the business, not only that of systems and standards. Showing up at strategic planning meetings and operational reviews is important. When they frame discussions around revenue, risk, and customer experience, leaders listen differently.

In companies where EA truly takes hold, executives use architectural ideas in everyday conversation. They refer to capabilities, compare current and future states, and debate the sequence of initiatives. Architecture stops being "the architects' thing" and becomes part of how everyone talks about change. That kind of alignment takes steady effort. Architects have to stay visible, build trust over time, and show real results people can point to.

A Working Definition of Success

Bringing these threads together, we can offer a practical definition: Enterprise Architecture is successful when it clearly improves strategic alignment, decision quality, speed of change, and organizational clarity.

Notice what's missing from that definition: documentation volume, framework compliance, or architectural purity. Those things can help. But they are not the goal.

Under methodologies such as the Kettera EA Methodology, you can spot success in clear, synced roadmaps, less duplicated work, smarter oversight, and a reliable rhythm that ties big-picture goals to day-to-day action. It works in the real world. You can see it happening and it lasts. It's built over time, not sparked by a one-off project.

The next chapter draws a line between a methodology and a framework. One is a concept while the other is a working system. Why does that difference matter? Because Enterprise Architecture only delivers real value when it runs on a clear, disciplined approach and not just a template.

CHAPTER 3

Methodology vs Framework

Enterprise Architecture uses terms that sound alike but aren't. *Framework* and *methodology* get treated similarly but they are different in important ways. A framework lays out structure. It organizes thinking, defines categories, and gives you a map of the territory. It does not give you a plan for execution. A methodology, on the other hand, demands action. It spells out steps, sequences decisions, assigns responsibility, and pushes work forward. One describes the landscape; the other tells you where to go.

Methodologies such as the Kettera Enterprise Architecture Methodology are built with that difference in mind. It is not a theoretical exercise or a reference model meant to sit untouched on a shelf. It is meant to be used. It favors doing over debating. It gives practitioners a clear place to begin and practical guidance for what comes next.

You should finish this chapter able to explain the gap between a framework and a methodology in plain language. You should also see why both matter. One shapes understanding while the other drives execution.

What a Framework Is

A framework provides structure. It lays out categories, perspectives, domains, and definitions that organize how people think about a problem. In other words, it clarifies *what* needs attention, but it does not always explain *how* to move from ideas to action.

Take a common enterprise architecture framework. It might break the organization into layers such as business, data, applications, and technology, and then specify the artifacts to create, the principles to follow, and the relationships that should connect those pieces. Those boundaries are useful and they keep conversations from drifting into vague abstractions. Instead of scattered opinions, the discussion has a shared map to work from.

However, a framework does not usually tell you how to run a quarterly review with business leaders. It does not spell out how to prioritize competing initiatives or how to reconcile two departments with conflicting technology choices. It offers guidance and shared vocabulary. It sets the stage.

Think of a framework as the blueprint within a building code. It defines what must be considered for a structure to be compliant: electrical systems, load-bearing walls, fire safety, plumbing. But it does not build the house. That work requires sequence, tools, coordination across trades, and real-time decisions.

Frameworks are therefore more descriptive than prescriptive. They outline the possibilities and encourage completeness. They create a common language across organizations and industries. On their own, they rarely create momentum.

What a Methodology Is

A methodology is about doing the work. It spells out the steps, the timing, the people involved, and the deliverables. A framework tells you what areas deserve attention; a methodology tells you what's happening next Tuesday, who's in the room, and what will be on the table by the end of the meeting. It turns ideas into a to-do list that is clear, dated and owned.

Artifacts get created on a schedule and refreshed on purpose, not by accident. It locks architecture into real-world cycles like planning, budgeting, and project approval gates, where money and priorities actually get decided. Miss that link and architecture drifts into theory.

Consider a simple example. A framework might say that current-state and future-state diagrams are required. A methodology goes further by answering the practical questions: When do we create these diagrams? Who signs off on them? How often are they reviewed, and what prompts an update? Where are they kept? How do they shape funding decisions?

It also defines accountability. Who owns the capability map? Who keeps the application portfolio up to date? When architectural standards clash with urgent business needs, who decides? This is more than structure, it defines how people act.

Why the Distinction Matters

Many organizations adopt a respected framework and assume they now "have Enterprise Architecture." They create documentation that matches the framework's categories. They may even train staff on the terminology. Yet six months later, little has changed in how decisions actually get made because a framework does not automatically change behavior.

Architecture shapes an organization only when it influences real decisions like what projects get funded, what systems are retired, what standards are enforced, and how trade-offs are resolved. Those decisions happen inside existing management processes: budgeting cycles, portfolio reviews, executive meetings, and project gates. A methodology ties architecture to the moments when those decisions are made. Without a methodology, architecture drifts toward documentation. Teams produce models because the framework recommends them. They catalog systems because the framework advises keeping an inventory. Over time, the artifact library grows, but its impact shrinks. To help avoid this, ask a hard question every time: how will this artifact actually change a decision? If you can't point to a clear shift in what someone will choose or do, it's time to rethink the artifact and maybe even scrap it.

Frameworks Provide Breadth; Methodologies Provide Motion

Frameworks are strong when it comes to scope. They help ensure no domain is ignored and promote balanced attention across business processes, information, applications, and infrastructure. Without them, architecture can drift into being either too technical or too abstract.

Methodologies, by contrast, create momentum. They move work forward through defined cycles and structured activities, set priorities, and shape the order of conversations. They also clarify when something is "good enough" to advance rather than debate endlessly.

Most enterprise architecture programs need both working together. One offers a complete view of the enterprise, while the other keeps efforts moving instead of stalling.

The most useful methodologies build on established frameworks and borrow what works from them. They do not introduce a new taxonomy or competing terminology. Instead, they concentrate on putting architecture into action and integrating it into the organization's regular decision-making cadence.

This distinction is subtle but powerful. Instead of asking, "Which framework do we follow?" methodologies like Kettera ask, "How do we consistently produce architecture that shapes planning and investment decisions?" That shift moves the conversation from classification to execution.

The Kettera Methodology as a Starting Point

The Kettera approach is meant to get you moving, not to lock you into a fixed playbook. It lays out key artifacts and shows how the work connects to business planning, but it never claims to be the last word. Built-in flexibility is part of the design.

Every company runs differently. One might stick to tight annual budgets. Another shifts money around every quarter. Some teams write long, formal documents. Others think better at a whiteboard and capture ideas in quick notes. Ignore those differences and what happens? The method won't stick, no matter how good it looks on paper.

Kettera starts with a clear foundation: defined architecture domains, a small set of artifacts people actually use, and a regular rhythm tied to portfolio and investment reviews. From that starting point, organizations shape it to fit their reality. Some introduce artifacts to meet regulatory demands. Others trim review steps to match a lean governance style. Many shift roadmap updates so they line up with fiscal planning cycles. What matters is customizing on purpose rather than letting things drift. Adjustments should be conscious choices that reflect real organizational needs and not the byproduct of inattention.

Practical Application: A Mid-Sized Enterprise

Imagine a mid-sized company with 2,000 employees and a growing portfolio of digital initiatives. The leadership team chooses a well-known architecture framework. It assigns a small EA team to put it in place. Over the next year, the team creates layered diagrams, principle documents, and a detailed application inventory. Despite all this work, project teams keep proposing redundant systems. Business units launch overlapping initiatives.

Technology standards are applied unevenly and the architecture team feels sidelined.

Picture the same company taking a method-first route. The enterprise architecture team locks in a quarterly architecture review that lines up directly with the investment committee's calendar. They pull out capability maps to spot where new proposals overlap or duplicate each other. They hold application portfolio reviews *before* the budget gets approved, not as a cleanup step after the money is gone. They refresh roadmaps ahead of planning meetings so leaders walk in informed, not guessing.

Documents and diagrams are no longer created just to check a box. They show up because they are needed in a real decision meeting. Each artifact has a job to do. The method spells out exactly where it plugs in and who uses it so there's no mystery and no busywork. The diagrams themselves aren't the key to success. The key is in the timing and context. What matters is when those visuals enter the conversation and how they shape choices.

Avoiding Methodology Rigidity

When every artifact is required in every situation, teams end up checking boxes instead of thinking. Discipline matters, but it should guide judgment and not choke it to where the methodology is too rigid.

Kettera argues for a minimum set of artifacts: only those that clearly shape decisions. If no one cites a document in roadmap debates, portfolio reviews, or approval meetings, it needs to be rethought. Otherwise, it's just paperwork. This keeps the approach sharp and useful.

Over time, the approach should shift. As architectural practices mature, teams often need fewer deep reviews because standards become second nature. During rapid growth or mergers, though, tighter oversight may make sense for a while. Methods should bend to fit the moment. The aim isn't to make every organization look the same. It's to keep strategy, investment, and execution moving in sync.

From Structure to Constraint

Understanding the difference between a framework and a methodology sets up a deeper conversation about constraints. Frameworks outline areas of concern. Methodologies guide action. Yet both operate within limits such as budget, culture, regulation, and technology. Architecture does not work in a vacuum. Even the most carefully designed methodology must respect the limits that shape what is realistic and practical. Some constraints come from outside; others are self-imposed through policy or history.

The next chapter digs into these limits using the framework of the "Three C's," breaking them down so you can see how they shape real decisions. What looks like restriction at first glance is actually structure. Enterprise Architecture depends on understanding where the edges are and choosing to work inside them. Treating constraints as part of the job is what keeps the practice steady and useful, and honest about what can and can't be done.

Key Takeaways

A framework gives you structure, shared language, and clear categories. It lays out what belongs in the conversation. A methodology, by contrast, sets the rhythm of the work and assigns responsibility. It explains how decisions get made and how architecture shapes those choices over time.

Think of frameworks as wide-angle views and methodologies as engines that keep things moving. Each has value, but only a methodology reliably ties architecture to planning and investment cycles. The Kettera EA Methodology is meant to be a practical place to begin so organizations can adapt the artifacts and meeting rhythm to fit their own culture.

CHAPTER 4

Architecture Constraints (Three C's)

Enterprise Architecture lives inside limits. Budgets get cut or delayed. Hiring stalls. Purchasing takes months. Teams lack key skills. People are maxed out. Old systems that are still running core processes can't just be switched off.

The Kettera approach faces these limits head-on. We call them the Three C's: **Capabilities, Capacity, and Capital**. Ignore even one of them and the plan falls apart. Respect all three, and you might actually ship something that works. An architecture that cannot operate within these three constraints simply isn't real. The Three C's aren't meant to box in your imagination. They anchor strategy in the real world, where plans have to be built, funded, and sustained.

Capabilities: What Can We Do?

Capabilities answer a simple question: *What is this organization able to do today?* This includes skills, tools, technologies, processes, and shared know-how. It is not about volume or scale. It's about what abilities exist. A company may have the capability to build cloud-native applications because it employs architects and engineers familiar with AWS or Azure. Another organization may have strong data analytics capability because it has data scientists and up-to-date reporting tools. A third might have deep regulatory compliance capability from years of operating in heavily regulated industries.

Capabilities are simply the things you can do. A leadership team might announce a shift to microservices. It sounds cutting-edge but do they actually employ engineers who grasp distributed systems? Do they have solid DevOps skills to support it? If the answer is no, the plan isn't strategy, it's wishful thinking.

Enterprise Architecture begins by mapping what the company can truly do today and not what it hopes to do next year. Take stock of technical know-how, core business strengths, day-to-day processes, and the tools that make everything run. A manufacturer may be great at fine-tuning production lines yet weak at building digital products. A bank could run airtight compliance automation but fall flat on customer experience design. Every company has its strengths and weaknesses.

Capabilities are not limited to people though. Tools count too. When a company standardizes on a specific integration platform, that platform itself becomes a capability. A data warehouse supported by mature governance processes is another example. On the other hand, dependence on spreadsheets and manual workflows points to limits in capability, even when skilled people are on staff.

An effective methodology asks architects to evaluate capabilities early, before defining target states. The goal is not to box in creative thinking, but to make the distance between today and tomorrow clear and measurable. With a grounded view of current strengths and constraints, decisions about what to pursue next become far more practical.

Capacity: How Much Can We Do?

If capabilities tell us *what* is possible, capacity tells us *how much* is possible.

An organization may have React development capability because it employs a skilled front-end engineer. That does not mean it can deliver five React projects at the same time. One developer cannot scale infinitely. The capability is there, but the capacity is not.

Capacity is about amounts. How many developers? How many servers? How much processing power? How many hours are available in a sprint? It looks at throughput and load. In infrastructure terms, capacity determines whether systems can handle 1,000 concurrent users or 100,000. In human terms, it shows whether teams are lightly loaded or constantly overwhelmed.

Many architecture failures come from mixing up capability and capacity. A leadership team might say, "We know how to build mobile apps." That may be true but if the mobile team is already tied up maintaining existing applications, new initiatives will stall. Capacity constraints create bottlenecks, even when the capability exists.

Think about a company replacing its old ERP system. The firm may employ database admins who know the system inside out. Those same people also keep daily operations running. Pull them onto a big overhaul project and something else may break. Leaders have to weigh the trade-off of progress on the new system versus stability in the old one.

People aren't the only limit. Network bandwidth, storage space, computing power, and even the fine print in vendor support contracts set hard caps. A data analytics push might call for near real-time reports, yet the current setup only handles overnight batch runs. When ambition moves faster than the infrastructure allows, capacity issues become visible.

Capacity determines how many efforts can move at once and how fast a long-term plan can unfold. When architects sketch out a 12- to 24-month roadmap, they have to base it on potential output. If we ignore those limits, the schedule may look great on paper and then fail due to day-to-day pressure.

A company running flat out with no slack leaves no room to try new ideas. Staff spend their days keeping the lights on instead of iterating towards the next improved state. In that situation, architects may need to suggest deeper fixes such as cutting back old systems, simplifying tools or trimming what no longer earns its keep to free up space for something better.

Capital: What Can We Fund?

Capital in this sense, means financial resources. It's what enables an organization to build new capabilities or expand capacity. With enough capital, a company can hire more developers, invest in new platforms, bring in consulting firms, or upgrade infrastructure.

Suppose an organization lacks cloud engineering capability. It can train current staff, hire experienced professionals, or work with a managed service provider. Each option costs money. The architectural gap isn't just technical; it's financial. Capital decides how fast that gap can close.

When a team reaches a limit, like there aren't enough data engineers to keep up with rising analytics needs, money can ease the pressure. You can hire more people. You can bring in automation tools to cut down repetitive work. You might even redesign workflows to remove bottlenecks. Funding stretches what the team can do; it turns hard limits into problems you can solve.

Other projects compete for the same funds. Leaders have to balance tech upgrades with sales goals, new market pushes, and keeping day-to-day operations steady. What happens to a architecture plan that assumes endless funding? It doesn't make it past review. An investment strategy that ignores financial limits won't survive executive scrutiny, no matter how good the design looks on paper.

For that reason, Enterprise Architecture has to present its recommendations in business language. Instead of saying, "We need to migrate to a new integration platform," the architect should spell out the financial logic behind the move. Which inefficiencies disappear? Which risks shrink? Where do new revenue streams open up? When leaders can see the value in concrete terms, funding decisions become far more straightforward.

Funding also comes with timing realities. Approval does not mean all the money arrives at once; large transformation efforts are often spread across several fiscal years. Architecture roadmaps need to match that rhythm. A phased modernization plan with clear value checkpoints is far more viable than a single sweeping overhaul that pushes beyond what the organization is willing, or able, to spend.

Here's a practical observation: money can address both capability and capacity gaps, but only when leaders spot those gaps early. Late discoveries drive up costs. Being explicit about constraints not only prevents sticker shock, it also makes funding requests more convincing.

The Interplay of the Three C's

Each "C" can be examined on its own, but in practice they rarely operate in isolation.

Imagine a healthcare organization overhauling its patient portal. It has strong web development capability but limited cybersecurity expertise. Its development team is already close to full capacity maintaining legacy systems. Meanwhile, capital is available because leadership has made digital engagement a priority.

Money can be directed toward bringing in cybersecurity experts or hiring an outside security firm. Either move adds the know-how that's missing. Teams can also free up breathing room by temporarily filling maintenance jobs so core staff aren't stretched thin. The skills gap shrinks, the workload lightens, and the project moves ahead with far less risk and far fewer surprises.

The Three C's form a simple decision matrix:

- No capability but enough capital? Buy it or build it.
- Capability in place but no capacity? Add resources or reset priorities.
- No capital? Rethink the scope, the sequence, or the timing.

This approach turns vague limits into clear options and helps us identify specific challenges when things aren't working out the way we expected. Rather than saying, "The team is not delivering," leaders can ask, "Is this a capability problem, a capacity problem, or a capital problem?"

Applying the Three C's in the Kettera Methodology

In the Kettera Enterprise Architecture Methodology, the Three C's are evaluated during early planning and revisited throughout roadmap development. Before defining the future state, architects document current capabilities across business and technical domains. They note both strengths and areas that need investment.

Next comes capacity analysis. Teams assess project pipelines, staffing levels, operational workloads, and infrastructure limits. This step helps prevent overcommitment and highlights where sequencing changes are needed.

Finally, capital alignment ensures that proposed initiatives match budget realities. Architects work with finance and executive leadership to confirm that transformation plans are financially viable. If not, the scope is adjusted.

Organizations that make a habit of reviewing the Three C's build real strategic instinct. They develop a clear picture of where they are strong, where they're stretched thin, what capabilities are missing, and how much time and money it will actually take to close the gaps. They stop operating on hunches and start working from evidence. The payoff is confidence rooted in facts, not optimism. That steady self-check, done consistently, sharpens judgment over time.

Key Takeaways

Enterprise Architecture operates within real-world limits. The Three C's: Capabilities, Capacity, and Capital, set those limits.

Capabilities define what the organization can actually do. Capacity sets the pace and volume of that work. Capital supplies the money needed to maintain, strengthen, or expand both. Ignore any one of these, and architecture drifts away from what the business can truly deliver.

When architects assess the Three C's directly, strategy stays tied to day-to-day operations. Instead of broad ambitions, they see specific gaps and clear investment choices. That shift sharpens roadmaps, makes funding conversations more concrete, and lowers the risk of stalled delivery.

With these constraints clear, the next step is to see how they fit into a structured method. The following chapter provides an overview of the Kettera Architecture Methodology in full and explains how these ideas come together in a practical, repeatable system for Enterprise Architecture.

CHAPTER 5

Kettera Architecture Methodology

We have covered the basics of Enterprise Architecture, spelled out what good results actually look like, separated methodology from framework, and explored the limits that shape every architectural choice. But ideas need to be applied and put into action. Architects sit in budget meetings, work through roadmap priorities, sketch diagrams that steer millions of dollars, and make calls that ripple across the business. Theory doesn't survive long in those rooms unless it can be used.

This book relies on the Kettera Architecture Methodology (KAM) but It is not the only way to practice Enterprise Architecture. Plenty of methods succeed when people apply them with focus and consistency. KAM simply gives us a stable point of reference, a shared structure we can return to again and again.

KAM gives us a common language and a practical system we

can apply directly without constant translation required. It lets us shift from principle to action without losing clarity or momentum. KAM is built for the real world. It starts from the fact that time is tight, priorities compete with each other, and most organizations are working with systems they already have in place. Instead of chasing perfection, it focuses on making steady progress in a clear direction.

Why a Named Methodology Matters

You may wonder why we need to name and formalize a methodology at all. If architecture is about aligning technology with business goals, can't we just do that without putting a label on the process?

In practice, not having a clear method leads to inconsistency. One architect creates detailed models; another relies on casual conversations. One team reviews every initiative; another steps in only after projects are already underway. Without a shared approach, architecture becomes personality-driven. The results depend on who happens to be in the room.

KAM solves this by laying out a repeatable way of working. It clarifies how architecture work begins, how artifacts are created, how decisions are documented, and how progress is tracked. That repeatability builds trust. Business leaders start to see what to expect from architecture. Technology teams understand when and how they'll be involved. That predictability cuts down on friction throughout the process.

This does not mean rigidity. The methodology offers structure, but it can and should be be adjusted based on an organization's size, maturity, and urgency. A regional nonprofit will not use KAM in the same way as a multinational bank. The principles stay the same but the depth and level of formality can differ.

The Core Structure of KAM

KAM rests on five tightly connected parts:

1. Business Alignment
2. Architectural Domains
3. Artifacts and Deliverables
4. Governance and Decision Flow
5. Cadence and Continuous Improvement

Each piece supports the others. This isn't a step-by-step program you march through and finish. It's a working system that is always moving, always interacting. If Business Alignment slips, decisions become unstable. If governance breaks down, delivery slows. Weakness in one area drags the rest down with it. When we strengthen a single component, the whole structure gets sturdier. Let one decay and cracks spread.

Business Alignment anchors the methodology. Architecture starts with the organization's goals, not with the latest technology trend. When a company aims to enter new markets, architectural priorities shift toward scalability, localization, and integration readiness. If leadership is focused on improving efficiency, the emphasis moves to automation, streamlining processes, and consolidating systems.

Alignment isn't something you check once and forget. Strategies evolve, and leadership teams reset priorities. KAM expects architects to reassess and confirm alignment on a regular basis instead of assuming last year's direction still holds.

The second component, **Architectural Domains**, breaks the enterprise into clear viewpoints that can be managed. This book explores both Business and Technical Architecture domains in depth including marketing, operations, finance, applications, data, integration, and platforms. Each domain must be addressed deliberately. A major initiative in one area should never move forward without recognizing how it affects the others.

Take the example of introducing a new customer relationship management (CRM) system. That choice goes far beyond selecting an application. It shapes data standards, integration approaches, business workflows, reporting models, and even team responsibilities. KAM pushes architects to evaluate these cross-domain impacts methodically rather than responding only after problems appear.

Artifacts That Drive Action

Methodology comes to life through deliverables. KAM defines a focused set of artifacts that guide decisions and support communication. These are not academic documents meant to sit in a repository. They are practical tools people actually use.

A capability map clarifies what the organization must be able to do, independent of specific systems. Current and future state diagrams show where the organization stands and where it plans to go. Roadmaps lay out the transition over 12 to 24 months.

Application and technology portfolios show what exists today and what is planned. Architecture Decision Records capture the reasoning behind major choices.

Consider a mid-sized healthcare provider planning to modernize its patient management systems. Without clear artifacts, discussions drift into vendor comparisons and long feature lists. With KAM, the conversation starts with capabilities: appointment scheduling, patient intake, billing, and clinical documentation. The current state diagram exposes redundant systems and manual workarounds. The future state sets integration standards and shared data models. A roadmap sequences changes to reduce disruption. Each decision is documented so future teams understand why certain paths were chosen.

Artifacts create shared understanding. They also provide continuity when staff changes. Architecture should not rely on institutional memory alone.

Governance Without Gridlock

"Governance" makes people nervous. Teams picture slow meetings, rigid rules, and review boards that kill momentum. In many companies, architecture oversight feels like a checkpoint designed to say no. KAM takes a different stance. It treats governance as a focused, structured conversation and not a bureaucratic wall.

In KAM, governance spells out how new work gets looked at, who needs a seat at the table, and what yardsticks are used. It draws clean lines around decision-making. What can a project team decide on its own? What needs a quick architectural review? What rises to the executive level? Clear answers will prevent conflicts later.

Consider a business unit that wants a niche analytics tool. In a loose setup, they buy it quietly and move on. After a few months, the tool won't connect cleanly with core systems, and the company realizes it's paying for features already covered elsewhere. Cleanup follows which is expensive, slow and frustrating. With KAM governance, that same proposal gets checked early against data rules, integration plans, and the broader tech roadmap before any contract is signed. The aim isn't to stifle creativity. It's to avoid messy duplication and scattered systems.

If standards repeatedly clash with how projects actually run, the standards need fixing, not the teams. KAM assumes the process can change. It's meant to help the organization move forward, not trap it in paperwork.

Cadence as a Discipline

Architecture initiatives often begin with energy and optimism. Workshops generate diagrams, ideas, and a sense of momentum. Then day-to-day operational pressures take over, and attention drifts. KAM reduces that pattern by establishing a clear, repeatable cadence.

Cadence is the steady rhythm of architectural work: quarterly roadmap reviews, monthly portfolio check-ins, recurring domain discussions, and scheduled governance sessions. These touchpoints keep architecture in view instead of letting it slip into the background. They also provide built-in opportunities to adjust direction before small issues grow.

You can think of cadence as the methodology's heartbeat. When the rhythm is broken, artifacts age quickly and alignment starts to erode. When it is kept constant, architecture stays tied to active initiatives and real decisions. Ideas move through a defined pipeline, priorities are revisited at the right moments, and progress becomes visible.

Cadence also reinforces accountability. When leaders know roadmap updates are due each quarter, talks about delays or scope changes stay fact-based, not reactive. Architecture shifts from an advisory role to an operational discipline woven into the organization's management cycle.

Flexibility Within Structure

People often worry that formal methods will kill creativity. KAM handles this by separating the core ideas from the way those ideas get carried out.

The core ideas stay steady: tie work to business goals, make information visible across teams, write down key choices, and review progress on a regular rhythm. Those don't change. What changes is how you put them into practice. A small startup might sketch simple diagrams and jot down short decision notes. A heavily regulated company, on the other hand, may need thick documentation and strict approval steps. Both approaches fit inside KAM. Enterprise Architecture has to match the size and experience level of the organization. Push too much process onto a tiny, fast-moving team and you'll get eye rolls and workarounds. Use too little structure in a large, complicated company and things fall apart fast.

KAM acts as a steady frame, the backbone that holds everything upright, while the surrounding practices stretch or tighten as needed. It can stay light and quick, or grow more formal and detailed. The frame holds while the details adjust.

At times, architects have to move fast and make provisional calls before all the analysis is in. KAM supports this reality by encouraging iteration: a decision can be logged as temporary and paired with a specific review date. By making that status visible to everyone, the team is less likely to let a "temporary" fix quietly turn into a permanent one.

KAM in Practice: A Scenario

To make this more concrete, consider a manufacturing company that wants better visibility across its supply chain. Leadership is aiming for real-time tracking of inventory across multiple warehouses and partners. Using KAM, the work starts with business alignment. The goal is clear: reduce stockouts by 20% and improve order fulfillment times. The team then looks at domain impacts. They review operations processes, data quality, supplier integrations, and platform scalability.

Several artifacts are created: a capability map that highlights inventory management and supplier collaboration; a current state diagram showing siloed systems; and a future-state architecture with centralized data and standardized APIs. An 18-month roadmap lays out warehouse upgrades, integration work, and analytics rollout.

Governance reviews make sure the chosen technologies fit data standards and the long-term platform direction. Regular meetings track progress and adjust timelines when needed. Key decisions such as choosing a specific integration platform are documented for future use.

KAM never tells you which vendor to pick or what software to buy. It doesn't hand down a preferred tool or lock you into a single option. What it does is shape the way you think through choices and measure their consequences. The focus stays on how decisions get made and how their ripple effects are judged, not on the decision itself. The approach builds guardrails around your thinking, without boxing you in. You get a clear, steady way to weigh trade-offs and look at impact, yet the final call remains yours. KAM brings rigor to the process while staying neutral on the outcome.

Why We Use KAM in This Book

In the chapters ahead, we use KAM as our running example. Capability maps, roadmaps, governance models, and innovation pipelines are all explained through this lens, so ideas connect instead of appearing as isolated cases. The goal is steady understanding built step by step, not a string of unrelated illustrations.

KAM is not the only valid path. You can adapt parts of it, blend it with what you already use, or line it up with other standards or industry-specific frameworks. The label matters far less than having a clear, repeatable system in place.

Key Takeaways

The Kettera Architecture Methodology provides a clear but flexible framework for Enterprise Architecture. It brings together business alignment, defined domains, practical artifacts, governance processes, and recurring cadence into one unified approach. It favors practicality over theory and progress over perfection.

By using KAM as our reference model, we gain a consistent way to examine Business Architecture, Technical Architecture, and the artifacts and rhythms that support them. In the next section of the book, we move from methodology to practice, starting with Business Architecture and a principle that supports the entire approach: organizational goals must come before technology goals.

Business Architecture

CHAPTER 6

Organizational Goals Over Technology Goals

Every new platform claims it will make things faster. Every new architecture says it will make you more flexible. Vendors talk about "transformation" like it's a switch you flip. Inside companies, excitement around the next new technology trend can be contagious but enterprise architecture doesn't start with tools or frameworks. It starts with a business need.

Business Architecture grounds a company in its reason for being and how it actually makes money. When technology becomes the starting point instead of the support system, results don't always align with what the business actually needs. Money gets poured into systems instead of business results. Wins are counted in go-lives rather than real impact. Before long, the company is busy but no closer to its goals. An organization can modernize its stack, migrate to the cloud, and automate workflows and still fall behind competitors who stayed focused on customers and value.

Enterprise architecture should keep asking the question: what problem are we solving, and for whom? If the answer centers on the technology itself, something has gone wrong. Tools are there to serve strategy, not the other way around.

This chapter lays out a straightforward but firm rule: business goals come before technology goals, every time. In the Kettera approach to Enterprise Architecture, the starting point is the organization's mission, its strategy, and the outcomes it can measure, not the tools, not the platforms and not the latest trend. Technology choices are made only after those aims are clear and agreed upon, and they are judged by how well they advance them.

Purpose-Driven Architecture

Every organization exists to accomplish something specific. It may aim to manufacture goods efficiently, deliver healthcare safely, provide financial services securely, or create innovative consumer products. That purpose defines its mission, guides its vision, and shapes the value it offers to customers. Enterprise Architecture has to start there.

When architects begin with mission and strategy, they ask different questions. Instead of "Which cloud provider should we standardize on?" they ask, "What capabilities do we need to strengthen to carry out our strategy?" Instead of "Should we adopt microservices?" they ask, "Where do we need speed, flexibility, or scale to support growth?" Technology decisions come later. They depend on the answers to those questions.

Consider a regional healthcare provider focused on improving patient outcomes while lowering costs. If its strategic objective is to reduce hospital readmissions by 15 percent over two years, architectural priorities shift. Investments might center on care coordination systems, patient engagement tools, and analytics to spot at-risk patients. Infrastructure modernization may still happen, but only if it supports that objective. The mission drives the architecture, not the other way around.

Purpose-led architecture changes how executives talk about technology. Architects don't bring in a standalone roadmap and hope it sticks. They connect every project to a clear business aim. Each system, upgrade, or platform shift points straight at what the company is trying to achieve. Why fund anything if it doesn't move the mission forward?

When leaders can trace a straight line from architecture work to company goals Enterprise architecture stops looking like a side advisory group and it becomes a core driver of results. It shifts from giving suggestions to helping deliver the plan itself, where strategy actually turns into action.

Value Creation Over Technical Elegance

Architects value clean designs, and engineers take pride in elegant code. Infrastructure teams aim for stable environments with clear standards. Technical excellence is admirable but not the objective.

If a system looks impressive but does nothing to increase revenue, cut costs, reduce risk, or improve the customer experience, its impact on the business is limited. By contrast, a modest solution that automates a manual finance task and saves millions each year can matter far more. Enterprise Architecture should judge success by business results, not by how refined the design appears.

Consider a company that pours resources into reshaping its internal applications around the latest architectural patterns. The result is modular, scalable, and technically striking. Yet the business process behind it is still slow and out of step with customer needs. Sales stall and customer churn rises. People inside the company may admire the architecture, but it hasn't moved the business forward.

Now picture a retailer that starts with a different priority: improving customer retention. The architecture team partners with marketing and operations to deliver better customer segmentation, real-time inventory visibility, and simpler returns. The technology behind the scenes may not win awards, but customer satisfaction climbs and repeat purchases grow. What matters is not how advanced the platform looks but the results it produces.

Enterprise architects must always ask: What value will this create? How will we measure it? If the answer is vague or not clearly tied to financial or operational results, the project likely needs rework.

Technology as an Enabler, Not an End State

Systems, platforms, integrations, and data models are there to support what the company actually needs to get done. They're tools, nothing more. When leaders start treating "modernization" as the finish line, they confuse motion with results. Busy does not equal better.

The Kettera approach makes this clear by putting business capabilities at the center. Start with what the organization must be able to do: manage orders, process claims, onboard customers, develop products. Those abilities define the business. The software comes later. Technology serves the capability; it doesn't set the agenda.

Picture a logistics firm trying to stay competitive. It decides that real-time shipment visibility is essential. That's the capability. To make it real, the company might rely on mobile apps, tracking sensors, integration layers, analytics dashboards, and solid data pipelines. The architecture work focuses on delivering that visibility consistently and at scale, day after day, shipment after shipment. If one integration tool gets swapped out next year, that's fine. The core capability stays in place. The tools can change without changing the business goal.

This way of thinking also cuts off endless upgrade cycles that don't make business sense. Moving to a new platform just because it's newer won't impress senior leaders. Moving because it speeds up product launches or trims infrastructure spending by 20 percent would be considered a concrete benefit. Architects are most effective when they shift the conversation from technical features to those business capabilities. Executives don't need a deep dive into every system detail; what they need is a clear view of how technology helps the organization execute its strategy.

Aligning Architecture to Business Outcomes and KPIs

Purpose and value must be measurable. Without clear outcomes and key performance indicators (KPIs), architectural alignment stays theoretical. Strong enterprise architecture connects each major investment to specific, visible business metrics.

Start with strategic objectives. If the organization aims to grow revenue by entering new markets, architecture initiatives might include scalable digital channels, localized compliance capabilities, and better partner integration. Each initiative should tie to revenue growth targets, faster time-to-market, or cost-of-entry measures. The relationship doesn't need to be simple, but it must hold up under scrutiny.

Suppose a financial services firm sets a goal to cut operational risk incidents by 30 percent. The architecture team might focus on improving data governance, modernizing access controls, and automating monitoring. Those initiatives should clearly reference the risk reduction target. Track progress not just through system deployment milestones, but through real drops in audit findings or security incidents.

This traceability builds discipline. When budgets tighten, as they inevitably do, leaders can assess which architectural initiatives most directly support defined outcomes. Projects without measurable alignment are harder to defend. That's a strength of a mature enterprise architecture practice.

The Kettera approach links every key artifact to real business results. Capability maps and roadmaps, for example, are not paperwork exercises; they point directly to measurable outcomes. These tools exist to move the business forward and every diagram should show how performance improves.

Avoiding Technology-Led Strategy Drift

Every year brings a fresh wave of tools, vendor pitches, and industry buzzwords. Artificial intelligence, blockchain, automation platforms. The names change, but the pattern doesn't. Each one is marketed as the next source of competitive advantage. Without clear guardrails, however, strategy can start bending toward whatever technology seems most exciting at the moment.

Technology-led strategy drift happens when curiosity and hype outweigh real business needs. A board member reads about a competitor's AI rollout, and suddenly every department is told to "add AI" to its plans. A vendor presents a tempting pricing package, and adoption begins before anyone has built a solid business case. Over time, almost without anyone noticing, funding and attention shift away from established priorities and toward the newest trend.

Governance processes should require a clear explanation of the business problem before approving major technology investments. Architecture review boards must ask important questions like: Which strategic objective does this support? What KPI will improve? What capability gap does it close? This does not mean ignoring innovation. Structured experimentation can be highly valuable. Pilot programs and innovation pipelines let the organization explore emerging technologies responsibly. The key is that exploration should inform strategy, not override it.

When enterprise architecture provides this discipline, it protects the organization from chasing trends. It keeps the focus on long-term goals, even while adapting to new opportunities.

Executive Sponsorship and Business Ownership

Enterprise architecture only stays focused on real company priorities when business leaders are clearly in charge of it. Treat it like a back-office IT activity and it will chase system uptime, tool upgrades, and internal efficiencies instead of growth or market results. Senior executives must visibly back the work, and business owners must be named and accountable for what it delivers.

Leaders on the business side decide what winning looks like. They own the revenue results, guard the budget, answer for regulatory demands, and stand behind promises made to customers. Architecture's role is to turn those expectations into practical capabilities and enabling technology. That relationship should be spelled out plainly, with no confusion about who owns outcomes.

Major architecture programs need a specific executive sponsor from the business. Governance groups must bring in operational and finance heads, not only CIOs and chief architects. Status updates should open with business results and only then cover technical progress. When a platform launches, the real test isn't just "Did it go live?" but "Is performance actually better?" If the metrics that matter to the business don't move, what exactly did we accomplish?

Shared accountability shifts how people act. Engineers and architects make different choices when they know success will be judged against revenue, cost, or customer measures. Executives lean in when they see architecture as a way to execute strategy, not a compliance ritual or paperwork exercise.

Kettera and similar methodologies weave architecture directly into the rhythm of enterprise planning and delivery cycles. It isn't a side exercise or afterthought It shapes how decisions get made and where money goes.

Key Takeaways

Organizational goals must always come before technology goals. Enterprise architecture starts with mission, vision, and strategy, not with platforms or design patterns. Technical excellence matters but only if it creates measurable business value.

Architecture should enable capabilities that drive revenue, lower cost, reduce risk, and improve customer experience. Every major initiative must link to clear outcomes and KPIs. Governance guardrails prevent drift toward trends, while executive sponsorship keeps accountability tied to business results.

When technology serves a clear purpose, architecture becomes a powerful engine for execution. When technology becomes the goal, architecture risks turning into an expensive distraction.

CHAPTER 7

Organization Strategy

Enterprise Architecture doesn't start with software. It starts with intent. Before anyone picks a tool or draws up a platform diagram, the organization has already decided where it wants to go. Maybe the goal is rapid growth. Maybe it's protecting market share. Maybe it's cutting costs or reinventing the business through new ideas. Those decisions, clear or half-spoken, set the tone for everything that comes next.

In the Kettera EA Methodology, Business Architecture holds the center of the discussion. It frames the bigger picture so technical choices don't drift. Strategy gives architecture its reason for being. Without it, design turns into a grab bag of personal preferences and favorite tools. With it, architecture becomes a practical way to get things done, an engine for carrying out the plan, not just documenting it. What happens when no one agrees on the destination? You get systems that look impressive but pull in different directions. Clear strategy keeps everyone building toward the same outcome.

This chapter looks at how strategic intent turns into architectural priorities in practice. We'll move through measurable objectives, competitive positioning, growth models, risk tolerance, capital allocation, and the fit between strategy and the operating model. Together, these factors shape how Enterprise Architecture is designed and governed, and why it takes the form it does. This is where strategy stops being abstract and starts affecting real decisions.

Strategic Objectives and Measurable Outcomes

A strategy that cannot be measured cannot guide architecture. Organizations often describe their direction in broad terms: "We want to grow," "We aim to lead in innovation," or "We intend to improve efficiency." These are starting points, not instructions for action. Enterprise Architecture needs clarity.

Consider an organization targeting 20 percent revenue growth over three years. That goal immediately shapes architecture decisions. Systems must scale without losing performance. Data platforms need to support advanced analytics to uncover new opportunities. Integration capabilities must enable quick onboarding of new partners or products. Investment moves toward scalability, automation, and customer insight instead of small, incremental cost cuts.

Now contrast that with a company focused on cost leadership. The architectural priority shifts to standardization, automation, and removing redundancy. Overlapping systems become candidates for consolidation. Custom development faces tighter review. KPI priorities shift as well. Unit cost, cycle time, and system utilization become guiding measures. Architecture teams evaluate platforms not for feature depth, but for operational efficiency and total cost of ownership.

Innovation-driven strategies create a different architectural profile. If time-to-market is the main differentiator, rigid change processes may be loosened in controlled ways. Teams provision sandbox environments. Modular architectures are favored over monolithic ones because they support experimentation without disrupting core operations. Technical debt may be tolerated temporarily because speed matters more.

Strategic goals have to show up in hard numbers inside the architecture. When a company says it wants a better customer experience, that promise should surface as concrete targets like uptime and clean, reliable data. What good is a customer-first slogan if the system crashes or serves up bad information? Those metrics are the scorecard.

Chasing growth in new regions changes the blueprint, too. Expansion demands built-in localization and clear compliance controls, both measurable and testable. If those pieces aren't delivered, the growth plan will stall.

Architecture isn't some abstract exercise. It's the machinery that turns strategy into working results. Done right, it converts big plans into delivered capabilities and measurable performance.

Competitive Positioning and Differentiation

Every organization competes on something. It could be price, speed, customer intimacy, product innovation, or some mix of these. Enterprise Architecture should strengthen that choice.

When speed is the promise, systems must move just as fast. Processes need tight integration, and real-time data flows aren't a nice-to-have; they're essential. Batch processing that slows things down can break the very commitment made to customers. In these settings, event-driven design, low-latency integrations, and highly available platforms aren't technical choices. They're business decisions.

Companies that stand out through customer experience tend to focus on unified data and visibility across channels. When marketing sees one version of the customer, service another, and billing a third, friction is inevitable, and customers will notice. Architecture has to deliver a single, trusted source of truth. That often means enterprise-wide data governance, master data management, and clear API standards so information stays consistent wherever it appears.

Cost leaders play a different game. They win through standard platforms and streamlined application portfolios. Here, the guiding principle is consistency: fewer technologies reduce training expenses, simplify maintenance, and strengthen vendor negotiations. Heavy customization works against a cost strategy, so it's handled as an exception and not the default.

Product innovation strategies need modularity and rapid prototyping ability. Architecture becomes a platform for experimentation. Open APIs, microservices, and cloud-native environments support continuous iteration. Governance still exists, but it is tuned to support rapid development while protecting core assets.

Competitive positioning shapes where complexity is allowed and where it is removed. Enterprise Architecture makes that trade-off clear.

Growth Strategy: Organic, Acquisition, and Geographic Expansion

The way a company expands, shapes the way its systems and structure need to change.

When growth happens organically, the same business just gets bigger. More users lead to more transactions. This leads to more pressure on the systems already in place. The real challenge becomes scale. Can your tech handle double or triple the traffic without doubling or tripling the bill? Manual work starts to break down under the strain, so automation steps in. Cloud infrastructure becomes especially appealing so you can flex up or down as demand shifts.

Acquisition-led growth is a different beast. Buy another company and you inherit its tech stack, its data definitions, and its ways of working. None of it was designed with your systems in mind. If you expect to keep buying companies, you need a repeatable way to plug them in. That usually means clear API standards, agreed rules for mapping data, and a simple decision process about what stays, what goes, and what gets connected, even if it's temporary. Without that discipline, every deal adds more unnecessary complexity.

One acquisition runs on one platform, the next on something totally different. Over time, you end up with a tangled mess of tools that don't talk to each other. Efficiency slips and leaders lose a clean view of the business. Enterprise Architecture has to prevent that drift by setting a clear integration plan. Many firms split this into two phases: what must connect on day one so the business keeps running, and what gets streamlined later to fit long-term goals.

Expanding into new geographies brings a new layer of regulatory and localization challenges. Each region has its own rules around data residency, language requirements, tax policy, and privacy standards. To keep up, the architecture needs configurable compliance settings and business rules that can adjust by market. A system built for a single country often strains under global pressure, while a setup that's too decentralized can splinter reporting across the enterprise.

Growth paths also shape how much technical standardization and oversight are needed. When leadership is clear about the expansion strategy early on, the architecture can be designed to support it instead of scrambling to catch up later.

Risk Tolerance and Regulatory Posture

Every organization has a risk appetite, whether formally defined or informally practiced. Some operate conservatively, prioritizing stability and compliance. Others accept risk in pursuit of speed or innovation. Architecture must reflect that stance.

In heavily regulated industries such as financial services,

healthcare and utilities, security architecture and compliance controls are foundational. Segregation of duties, encryption standards, audit logging, and disaster recovery capabilities are not optional. They are structural requirements. Redundancy is built into systems because downtime brings legal and reputational consequences.

Organizations with higher risk tolerance may prioritize speed over redundancy. They may accept shorter testing cycles or phased rollouts that expose limited user groups to early releases. That does not imply recklessness; it reflects a calculated trade-off. In these environments, architecture governance focuses on monitoring and rapid fixes instead of exhaustive pre-deployment validation.

Risk appetite shapes how a company picks its vendors. A careful organization often steers clear of relying on just one supplier and lines up backup providers for any service it can't afford to lose. A bolder rival may do the opposite, choosing to bundle everything with one key partner to get better pricing and smoother integration, even if that means deeper dependence.

Enterprise Architecture has to turn abstract talk about "risk" into clear design choices people can see and manage. How often are backups run? Who gets access to what? How is the network carved up? What kind of cloud setup is used? These decisions quietly expose how much risk the organization is truly willing to live with. After all, what's the point of claiming to be risk-aware if the systems tell a different story?

Investment Horizon and Capital Allocation

Strategy is not just about choosing a direction; it is also about deciding when results must appear. Some organizations prioritize quarterly earnings, while others plan around a five- or even ten-year horizon. That time frame shapes how much architectural runway they build and how aggressively they pursue modernization.

When profitability in the near term drives decisions, large transformation programs are often postponed. Teams favor incremental gains instead of broad redesigns, and technical debt can pile up if cleanup efforts lack an immediate payoff. In these settings, architecture groups have to master phased modernization, upgrading piece by piece while keeping the business running without disruption.

Organizations that think long term tend to commit funding to foundational capabilities. They invest steadily in data platforms, enterprise integration layers, and legacy system renewal. Returns may take years to materialize, yet leadership accepts the delay because it supports larger ambitions. As a result, architecture roadmaps stretch further into the future and are less tied to quarterly reporting pressures.

How money is distributed also shapes experimentation. Setting aside part of the budget for research and development gives architecture teams room to run pilots and proofs of concept without putting core systems at risk. When no such funding exists, innovation ends up competing with day-to-day operations, and stability usually wins.

Understanding the investment horizon helps prevent impractical architecture plans. You can't deliver a ten-year transformation on a one-year budget.

Operating Model Alignment

The way a company is set up, whether power sits in the center or is pushed out to the edges, with plenty of mixed versions in between, has a direct impact on how architecture is governed. The operating model isn't just org-chart trivia; it determines who gets to decide, how standards stick, and how technology choices play out across the business.

Centralized setups put most decision authority at the enterprise level. Teams use shared platforms and the same standards apply everywhere. Enterprise Architecture steps in as the rule-setter, choosing core technologies and defining how systems connect across business units. This can cut down on repeat work and keep things consistent across the company. It can also box in local teams that need room to move. The trade-off is control and uniformity on one side, and flexibility and speed on the other.

Decentralized organizations give business units or regions wide latitude. Teams choose their own systems and processes to suit local demands, while architecture sets only the minimum integration rules and shared data definitions. Rather than dictating every platform choice, it aims for alignment without locking everyone into the same mold.

In hybrid models, a centralized core sits alongside decentralized innovation at the edges. Finance and HR systems might be standardized across the enterprise, even as customer-facing apps differ by market segment. Governance follows suit:

enterprise standards cover foundational capabilities, and domain teams retain freedom where differentiation matters.

Challenges are introduced when the operating model and architecture pull in different directions. Push a centralized blueprint into a highly autonomous culture and it will be ignored. Take a hands-off approach in a tightly controlled company, and inconsistency and risk will creep in. Enterprise Architecture has to match the way authority and accountability actually work.

Bringing Strategy into Architectural Practice

The Kettera EA Methodology puts strong emphasis on traceability. Every architectural initiative should clearly link to a strategic objective. If a proposed system upgrade cannot be tied to growth, efficiency, risk reduction, or differentiation, its priority should be reconsidered.

In practice, this often starts with strategy mapping workshops. Executives define objectives and measurable outcomes. Architects then turn these into capability needs and architectural principles. From there, roadmaps and investment portfolios are brought into alignment. This creates a feedback loop: strategy shapes architecture, and architectural constraints help refine the strategy. When this discipline is missing, architecture slides into chasing cleaner code and tighter systems while losing sight of what actually helps the business.

Key Takeaways

Enterprise Architecture sits inside the broader organization strategy. Strategic goals need to be turned into architectural outcomes you can actually measure. How a company competes signals where systems should shine and where consistency matters more than uniqueness. Plans for growth shape decisions about integration, scalability, and governance. A company's appetite for risk guides how security and compliance are designed. The time horizon for investment affects how quickly modernization happens. And the chosen operating model influences governance design and who owns which platforms.

In the next chapter, we shift from enterprise-wide strategy to specific business areas, starting with Marketing. Seeing how strategic intent shapes marketing capabilities offers a clear, practical example of Business Architecture in action.

CHAPTER 8

Marketing Domains

Marketing is the company's front door. It's the place where plans collide with real customers and promises either hold up or fall apart. In Business Architecture, marketing domains spell out how a company finds its buyers, gets in front of them, convinces them, turns interest into sales, and keeps them coming back. These choices drive technology stacks, shape data models, determine integration patterns, and set the rules for oversight and control.

In the Kettera EA Methodology, marketing decisions form part of the enterprise's structural blueprint. Decisions about customer segments, channel mix, experience design, brand position, performance analytics, and compliance duties directly influence architectural needs. Systems must reflect who the business serves, otherwise friction develops between the marketing processes and the technology. Ignore marketing domains and the result is scattered tools, duplicate data, and muddled accountability. Include them and the enterprise gains focus, leverage, and room to grow.

Customer Segmentation and Capability Alignment

Segmentation sits at the heart of every marketing model. Global enterprises do not buy like small businesses, and small businesses do not behave like individual consumers. Offerings tailored to a specific industry operate differently from those aimed at a new region. These may look like commercial distinctions, but they carry real architectural implications.

Take a company that sells to both multinational firms and individual consumers. Enterprise customers need contract management, tiered pricing, account hierarchies, and dedicated support processes. Consumers, by contrast, want fast checkout, self-service tools, and straightforward subscriptions. Supporting both groups means building distinct capabilities, and often configuring systems in very different ways, even if the product itself is the same.

Customer segmentation shapes data models from the start. Enterprise accounts often require more hierarchal relationships, multiple billing contacts, negotiated pricing tiers, and compliance attributes. Consumer records are flatter but demand high-volume identity management and behavioral tracking. A single CRM schema rarely supports both without thoughtful design. Enterprise Architects must decide whether to extend a unified model or separate domains with controlled integration.

Segmentation also influences personalization engines and analytics. A marketing team targeting mid-market firms may prioritize industry signals and firmographic data. A consumer marketing team is more likely to focus on browsing history, clickstream events, and purchase frequency. Those priorities guide the choice of customer data platforms (CDPs), event tracking standards, and machine learning models.

A consumer push can spark millions of clicks, views, and replies in a single afternoon. An enterprise push looks quieter, with far fewer names in the pipeline, yet each one demands serious time, custom follow-up, and a sales cycle that drags on for months. The same system has to handle both scenarios.

When everyone agrees on who the audience is and treats that choice as a core design rule, the tech stack lines up around it. One group grabs a tool built for big enterprise pursuits, another installs something aimed at mass-market blasts, and suddenly nothing fits together. Integration turns into a patch job instead of a plan.

Go-to-Market Model and Channel Architecture

A go-to-market model defines how a product travels from concept to cash. Whether a company relies on direct sales, partners, digital self-service, marketplaces, or field marketing, each path brings its own architectural needs. The Enterprise Architect must make sure these channels work together smoothly instead of turning into isolated systems.

In a direct sales setup, CRM must connect with quoting, pricing, and contract tools. Sales teams expect real-time access to inventory levels, pricing logic, and complete customer history. To make that possible, a CRM typically connects through APIs to ERP, order management, and billing systems, creating a dependable flow of data across platforms.

Partner ecosystems add another layer of difficulty. Distributors and resellers often need portal access, delegated administration, shared pricing models, and commission tracking. Identity management grows more complex because external users require secure access without being classified as employees. That means role-based permissions, identity federation, and detailed audit logs are mandatory.

Selling through e-commerce sites and marketplaces shifts the architectural focus again. These channels rely on highly available web platforms, scalable checkout flows, fraud prevention tools, and integrated payment gateways. Order management and fulfillment systems must sync in near real time. Pricing engines need to handle promotions, bundled offers, discounts, and dynamic price changes. When a company sells both through its own storefront and third-party marketplaces, keeping pricing and inventory aligned across platforms becomes a serious integration test.

Field marketing adds another dimension. Events, trade shows, and local campaigns bring in leads that need to flow smoothly into central systems. Lead capture tools, mobile apps, and event platforms should sync with CRM and marketing automation systems. When those integrations are fragile, data shows up late, or not at all.

The core architectural principle is consistency across channels. A customer who starts in an online store and later talks to a sales rep should not be treated as two different people. Multi-channel consistency depends on shared data services, standard APIs, and unified pricing logic. This is what determines whether customers experience clarity or confusion.

Customer Experience Strategy and Journey Mapping

Marketing teams now structure their work around the customer journey: awareness, consideration, purchase, onboarding, retention, and advocacy. This journey shapes how the whole company coordinates its work.

Campaign tools have to connect smoothly with ad networks, social platforms, email programs, and analytics dashboards. Tracking tags and event feeds record what people click, watch, and ignore. That information moves into data warehouses or customer data platforms so teams can group audiences and aim messages precisely. When each channel measures things differently, reports stop making sense, and decisions start to become unclear.

Prospects browse websites, attend webinars, watch demos, and talk with sales reps. All of those interactions need to show up in both marketing automation tools and the CRM. Lead scores only work if the underlying activity data arrives clean and complete. Sales needs a clear view of what someone has read or downloaded before picking up the phone. Waiting days for systems to sync is not good enough; updates have to happen fast, almost as they occur.

Orders must pass through order management systems, payment processors, tax engines, and fulfillment platforms. The promise marketing makes, “buy in three clicks” or “get started instantly”, has to match what operations can actually deliver. What happens if a customer pays and then waits days for access?

If onboarding tools can't activate the service right away, the brand's big promise falls apart at the exact moment it matters most.

Retention and loyalty efforts typically run on analytics and automation. Triggered campaigns, renewal alerts, and upsell promotions all rely on clean contract records and clear behavioral data. When systems don't talk to each other, even one missed renewal notice can mean revenue simply disappears.

That's why journey mapping acts as a coordination plan. Enterprise Architecture aligns APIs, automation tools, data flows, and workflow engines so each stage works as intended. It pushes teams to answer practical questions: Where is customer identity first created? How is consent captured and stored? Which platform defines the official customer status? The responses shape the integration design and determine how reliably the entire experience runs.

Brand Positioning and Digital Platform Requirements

Brand positioning is more than a marketing slogan. It sets clear performance and capability expectations that technology must meet.

An organization that promises speed must invest in low-latency platforms, optimized content delivery networks, and automated fulfillment. Performance service-level agreements become part of the brand promise. Slow page loads or delayed confirmations damage credibility.

A premium service brand calls for personalization, concierge-style support workflows, and unified customer profiles. Customer service representatives need to see complete interaction histories. Digital platforms should enable tailored content and targeted offers. Without integrated data, personalization turns into generic messaging.

A company that competes on low cost has to build everything around doing more with less. Processes should be optimized, repeatable, and heavily automated so day-to-day expenses stay down. If your marketing tools depend on constant hands-on work, they clash with that strategy.

Brands built on innovation play a different game. They need room to test ideas quickly and roll out changes without breaking what already works. Tools like feature flags and solid A/B testing setups make that possible, letting teams experiment safely and often. The underlying architecture has to expect change, and make space for it, without turning into chaos.

Marketing Data and Analytics Dependencies

Marketing runs on data. Measuring campaigns, building attribution models, and generating customer insight all hinge on information that is accurate and up to date. The underlying data architecture has to support this work without introducing duplication or conflicting records.

Attribution, whether first-touch, last-touch, or multi-touch, depends on event-level tracking across channels. Web visits, email opens, ad clicks, sales outreach, and completed purchases need to connect back to a single customer profile. Resolving that identity is no small task; it involves matching algorithms, master data management (MDM), and tight data quality controls. This is where many teams discover how messy their data really is.

Analytics and consent management are closely linked. When a customer opts out of tracking, related event data must be suppressed or anonymized, and every connected system has to respect that choice. This only works when consent repositories are directly integrated with marketing platforms and enforcement rules are applied consistently.

Most organizations rely on a data warehouse or lakehouse as the analytical backbone. Dashboards and predictive models pull from these environments. Problems arise when operational systems fail to emit standardized events or when departments define metrics differently; reports get challenged, numbers stop lining up, and confidence in the data starts to slip.

Enterprise Architecture defines standard data models, integration standards, and data governance policies. It makes clear which system serves as the system of record for customer attributes, campaign performance, and revenue recognition. Without that clarity, marketing analytics turns into spreadsheet reconciliation.

MarTech Stack Governance and Rationalization

Marketing tech stacks don't stay small for long. One group

brings in an email automation platform to speed up campaigns. Another team signs up for a different tool to run social channels. A regional office adds its own CRM plug-in to handle local needs. The collection swells and costs creep up. Connecting everything becomes a headache.

Every new application adds another layer of technical baggage: APIs to wire up, unique login systems, data sync rules, and fresh security risks. Customer records start to multiply across systems. Integration workflows grow brittle so that one small change can break multiple connections. Licensing bills rise, but that's only part of it. Support demands more hours and compliance reviews take longer. Hidden inefficiencies and expenses add up.

Enterprise Architecture serves as the guardrail. Innovation isn't shut down; it's just steered in a workable direction. Architects review proposed tools before contracts are signed, checking how well they connect, whether they follow shared data standards, and how strong their security really is. They lay out reference designs for marketing automation, customer data platforms, CRM add-ons, and analytics tools. Rationalization efforts then spotlight overlap and push teams to combine or retire systems when it makes sense.

Standardization simplifies the plumbing and keeps information cleaner across platforms but security improves as well. Marketing systems handle personal data every day, making them prime targets for attackers. Central oversight makes sure encryption rules, access permissions, audit trails, and continuous monitoring are applied the same way everywhere. Governance isn't about standing still; it's about changing on purpose. Marketing should absolutely experiment, but within clear architectural guardrails.

Regulatory, Privacy, and Consent Implications

Marketing often relies on tracking, profiling, and cross-border campaigns. Laws such as GDPR, CCPA, and other regional privacy rules set strict requirements for how data is collected, stored, and used. These rules directly shape system architecture.

Consent must be explicit, recorded, and easy to retrieve. Systems need to support detailed consent choices, including email marketing, SMS, profiling, and third-party sharing. If a customer withdraws consent, updates must spread across all marketing platforms immediately.

Data residency rules often demand that personal information stay inside certain national or regional borders. Global marketing efforts have to respect those lines on the map. Where you host your cloud systems and how you handle backups are shaped by these limits, and the wrong setup can create headaches fast.

Audit trails must also be considered. Companies have to prove they're following the rules, not just claim they are. That requires detailed records of when customers changed their consent and who looked at their data. Choices about logging systems, how long records are stored, and who gets access end up doing the heavy lifting for compliance, and saving you during audits.

Regulators issue fines and customers lose trust when compliance breaks down. Who wants to explain a preventable privacy failure to the board? Effective Enterprise Architecture builds privacy into marketing platforms from day one.

Key Takeaways

Marketing domains shape much more than campaigns. They influence how data is structured, how systems connect, how platforms scale, and how identity and compliance are handled. The way you segment customers affects which capabilities you build and how systems are designed to support them. Go-to-market choices shape channel structures and integration patterns, while customer journeys demand coordination across a web of enterprise systems.

Brand positioning carries technical implications. It sets expectations for speed, depth of personalization, and how quickly content can be updated. Analytics needs drive decisions about master data, event tracking rules, and governance practices. At the same time, the steady growth of MarTech tools requires tight oversight to keep complexity and spending in check. Regulatory requirements add yet another dimension, weaving privacy and consent directly into the architecture and that often means design decisions get legal input earlier than teams expect.

When Enterprise Architects work closely with marketing domains, they create alignment across channels and consistency in customer interactions. That alignment makes it possible to deliver internally on the promises made to customers.

CHAPTER 9

Operations Domains

Strategy sounds impressive but operations is what turns those slides into shipped products and paid invoices. Marketing can shape the promise and executives can chart the course, but operations makes sure orders go out, suppliers get paid, staff are scheduled, and customers get help when things break. In Enterprise Architecture, operations isn't just a stack of back-office tools. It's the deliberate design of how work actually moves and how information, materials, cash, and decisions travel across the company from start to finish.

The Kettera EA Methodology starts Business Architecture by asking a basic question: how does this organization plan to run? Only after that answer is clear do we decide which applications, integrations, data structures, and platforms are needed to support it. The operations domain gives an honest picture of that intent because it shows how the company behaves day to day. Marketing can describe ambition and vision; operations shows whether the organization has the discipline to follow through.

Operating Model and Process Standardization

Every enterprise makes a choice about how it runs. Some concentrate decision-making and process control at the center; others give regions or business units wide latitude. In some companies, shared services handle most operational work. In others, those functions sit inside each unit. These aren't abstract management theories, they shape very real technology decisions.

Take a highly centralized company with global process owners. It will probably aim for one consolidated ERP system, with common workflows for procurement, inventory, and financial close. Master data is tightly controlled, and variation in process is kept to a minimum. Because consistent data is non-negotiable, modules are deeply integrated. Consolidating platforms becomes a priority, cutting duplication and making reporting far simpler.

Now picture a decentralized enterprise made up of semi-autonomous divisions. One division might run its own ERP, tailor processes to local market conditions, and choose niche tools that fit its needs. In that setting, the architecture typically relies on integration layers that link different systems while preserving local freedom. Instead of mandating a single way of working, architects concentrate on interoperability standards and shared data definitions at key points, like consolidated financial reporting where the numbers have to add up, no matter how independent the divisions feel.

Shared services models sit somewhere between these extremes. Finance, HR, or procurement may be centralized while manufacturing and distribution remain distributed. This hybrid model requires clear boundaries in the architecture. Core transactional systems must support standardized services, while interfaces allow for local operational differences. The architecture should reflect where standardization is non-negotiable and where variation is intentional.

Ignore the operating model, and you'll design the wrong systems. An architect who understands it can shape ERP design, workflow automation, and integration strategy to match how the business actually runs.

End-to-End Process Architecture

Think of architecture the context of value streams. Consider order-to-cash. An order comes in. Stock gets checked. Credit gets cleared and goods ship. The invoice goes out and cash lands. It's simple in concept but can lead to complicated systems. CRM, ERP, warehouse tools, shipping platforms, billing engines are all influencing the same journey. Build them as silos and you invite delays and bad data. Design them as one connected flow with clear data owners and tight integrations, and work moves faster.

Ownership must match the work. If Finance controls invoicing and revenue, then finance systems are the final word on those figures. If Operations runs fulfillment, then it make sense that shipment status lives in warehouse or plant systems. If you blur the lines you get endless reconciliations and dueling reports.

Plan-to-produce ties together forecasting, scheduling, materials planning, and the shop floor itself. Digital plans meet physical output. That link has to be well defined. Planning tools and manufacturing systems often need real-time updates. The architecture should handle event-driven triggers, manage exceptions, track performance, and keep the whole value stream in sync, from forecast to finished goods.

When Enterprise Architects explicitly map these end-to-end processes, the integration patterns start to reveal themselves. They can standardize data models around shared entities like customer, supplier, employee, and product, instead of defining them differently across systems. In this way, the architecture reflects how the business actually operates, instead of breaking it into disconnected pieces.

Operational Scalability and Throughput

Some organizations handle thousands of transactions a day; others handle millions an hour. Volume changes the architecture. A seasonal retailer preparing for peak holiday demand has very different infrastructure needs than a regional manufacturer with steady output. Architecture must reflect transaction volume, geographic distribution, and expected growth as core design inputs.

High-throughput environments require systems that scale horizontally and bounce back quickly from failure. Cloud-native architectures, distributed databases, and automated scaling tools become practical necessities rather than technical luxuries. Performance standards should be defined in business terms: maximum order processing time, acceptable latency for inventory updates, and batch windows for financial close.

Geographic distribution adds another layer of complexity. A

global enterprise may need data residency compliance in multiple countries, regional data centers to reduce latency, and follow-the-sun support models. Integration architectures must support asynchronous communication where network reliability varies. Batch processing in one region can clash with real-time demands in another.

Growth plans also shape automation strategy. If the organization expects to double revenue in three years without doubling headcount, process automation becomes a requirement. Workflow engines, self-service portals, and straight-through processing become strategic enablers. Architects must plan for not only current capacity but realistic expansion scenarios. If they don't, success becomes the bottleneck.

Supply Chain and Vendor Ecosystem Integration

No company operates in a vacuum. Every business relies on a web of suppliers, carriers, distributors, and contract manufacturers to get work done. That wider network shapes how systems must be designed. Internal apps aren't enough; the architecture has to connect smoothly with outside partners using clear B2B integration patterns.

Electronic Data Interchange (EDI) is still alive and well, especially in retail and manufacturing. Many companies depend on it to move purchase orders and invoices back and forth. At the same time, newer systems offer APIs that deliver live order status, shipment tracking, and up-to-the-minute inventory data. A supplier might get an automated order through EDI, then log into a portal or call an API to check demand forecasts. Both old and new approaches have to work side by side. Architects can't pick one and ignore the other.

Logistics partners add more moving parts. A warehouse system may have to keep inventory counts aligned with several outside fulfillment centers at once. Transportation platforms need to send and receive shipment details, proof of delivery, and status updates. Instead of waiting for nightly batch files, event-driven designs push out changes as they happen so everyone sees the same picture quickly.

Disruptions are inevitable. A supplier's system will crash. A port closes. A truck never arrives. What happens then? Systems should be able to shift orders to backup vendors, apply flexible sourcing rules, and show exactly what inventory is in transit at any moment. Real-time dashboards can flag rising risks before they become full-blown crises. Strong enterprise architecture makes that possible by building in supplier diversity, fast rerouting, and clear visibility across the entire network, not just inside company walls.

In the operations domain, integration architecture goes far beyond simple connectivity. It shapes how trust is built, how accountability is maintained, and how work continues smoothly across organizational boundaries and that's often where things fall apart if the design is weak.

Service Delivery and Support Models

Operational domains go beyond manufacturing and procurement. Service delivery, whether field service, customer support, or managed services, brings its own architectural needs.

The systems behind these models must connect customer data, work orders, knowledge assets, and performance metrics. Field service teams need strong mobile support. Technicians require access to work orders, service history, inventory levels, and technical documentation while on-site. Offline access can be essential in remote areas. Integration among CRM, asset management, and scheduling systems must be seamless to prevent duplicate data entry and scheduling conflicts.

Customer support operations rely heavily on ticketing platforms, knowledge bases, and communication channels such as chat, email, and voice. The architecture should bring these interactions together under a single customer record. If a customer reaches out after placing an order, the agent should see the order status right away. That visibility requires integration between CRM and operational systems.

Managed services come with constant SLA oversight. Teams have to automatically capture and report key numbers like uptime and response speed, along with how quickly issues get fixed. Monitoring tools feed this data into dashboards, and in some setups, straight into invoicing platforms. The system design must guarantee that SLA math is accurate, traceable, and easy to audit, especially when missed targets trigger financial penalties.

The service delivery setup connects front-end customer platforms with the core operational systems behind the scenes. It turns raw performance data into something customers actually experience.

Operational Risk and Business Continuity

When manufacturing stops, revenue stops with it. If a warehouse goes down, thousands of shipments can be delayed. Heavy reliance on a few highly specialized employees creates single points of failure, something many organizations discover too late. Enterprise Architecture needs to account for these operational realities from the start.

Recovery tiers should match actual business impact. Real-time failover isn't necessary for every system, but it is for some. Platforms that handle orders or financial transactions may require near-zero recovery time objectives, while internal reporting tools can often withstand longer interruptions. Decisions about redundancy, data replication, and backup schedules should align directly with what the business can and cannot afford to have offline.

Incident management systems add another layer of resilience. Monitoring tools should spot failures quickly and send alerts with clear paths for who gets notified next. Connecting monitoring platforms with ticketing systems can automatically create and track incidents. When disruptions happen, response processes should be written down and practiced.

Business continuity planning goes beyond technology. Staff coverage, backup facilities, and supplier backups all connect to system design. An architect who understands operational risk will design systems that fail gracefully across all of these concerns instead of collapsing completely.

Automation and Process Optimization Strategy

Companies often launch lean efforts or digital overhauls to cut waste and work smarter. Automation usually sits at the center of those plans and it shouldn't be bolted on as an afterthought. It has to fit the larger system design.

Robotic Process Automation delivers fast results. It handles repetitive clicks and data transfers across older systems with impressive speed. However, problems arise when teams depend on it too heavily, small changes in the underlying software can break dozens of bots overnight. RPA works best as a bridge solution, not the destination. Pair it with durable API connections and a redesigned workflow that removes the clunky steps in the first place.

Workflow platforms offer a steadier path forward. They map out each step clearly and enforce business rules automatically, cutting down on human touchpoints and making audits far easier. Their success hinges on clean, organized data, as messy inputs will sabotage even the best-designed process.

Artificial intelligence and advanced analytics now shape day-to-day operations. Predictive models flag equipment likely to fail before it actually does. Forecasting tools help teams set inventory at the right levels so cash isn't trapped on shelves. These systems rely on strong data pipelines and firm oversight. AI without disciplined data practices turns into noise and questionable conclusions.

Automation needs oversight. Governance should spell out who can change workflows, how bots and agents are tracked, and what happens when something breaks. Efficiency matters, but without clear guardrails, automation can create risks just as quickly as it removes them.

Key Takeaways

Operations domains define how work gets done. The operating model, centralized, decentralized, or hybrid, sets the tone for system consolidation and process standardization. End-to-end value streams like order-to-cash and plan-to-produce should anchor architectural boundaries and integration patterns.

Scalability, geographic footprint, and growth plans shape platform design and automation strategy. External ecosystems, including suppliers, logistics partners, and service providers, push architecture beyond internal walls and call for resilient integration methods. Service delivery models tie operational systems directly to the customer experience.

Operational risk and automation don't manage themselves. Business continuity plans, incident response, workflow automation, and AI-powered optimization aren't side projects. They belong at the core of Business Architecture and need to be included in the way business actually runs day to day.

CHAPTER 10

Finance Domains

Most people see finance as the rule enforcer or the “bank”. In Enterprise Architecture, it plays a much bigger role. Financial decisions influence both business and technology strategy through proper management of our scarce resources. The company’s financial stance sets the tone such as cautious and protective, bold and fast-moving, or somewhere in the middle.

Inside the Kettera Enterprise Architecture Methodology, Business Architecture starts with strategy and goals. The finance domains take those ambitions and turn them into numbers, limits, and funding logic. Strategy points to the destination while finance decides how fast we can move, and how hard leadership is ready to push the engine.

Financial Strategy and Capital Allocation

A company’s financial strategy shapes how bold its

architecture can be. An organization chasing rapid growth, expanding market share, or buying competitors will accept different technology risks than one focused on steady margins and dependable dividends. Architects need clarity on what leadership values most: reinvesting profits, returning cash to shareholders, paying down debt, or expanding through acquisitions. Those priorities set the pace for modernization and determine how much technical debt the business is willing to carry.

The contrast becomes clear when you look at two different companies. A venture-backed tech firm racing for market dominance might delay system clean-up to move faster. In that setting, architecture choices favor scalability and quick feature releases, even if cost efficiency is postponed. A mature public company focused on stable earnings, by contrast, may demand careful ROI projections before approving platform upgrades. It is also more likely to push back on replacing legacy systems unless the payback period is obvious and near term.

Where money is directed tells architects where to focus. When the board commits major funding to expand digital channels, roadmaps should tilt toward customer-facing platforms and stronger integration. If budgets are tight and the mandate is cost reduction, consolidation and simplification rise to the top of the list.

Technical debt is another area shaped by financial posture. Organizations that invest heavily in innovation may tolerate higher levels of technical debt in non-differentiating areas. Others may push for faster cleanup to lower operational risk. Architects must explain technical debt in financial terms such as cost of delay, risk exposure and maintenance burden, so finance leaders can assess it alongside other investment decisions.

Budgeting Models and Funding Mechanisms

Money flow shapes outcomes more than raw dollars do. Big companies love annual budgets, but that once-a-year ritual can choke long, complex change programs. If approval comes only every 12 months, your architecture plan has to march to that drum. Work that really needs 24 to 36 months of steady investment gets sliced into yearly chunks just to pass the gate, even when that slicing makes little technical sense.

Quarterly reforecasting changes the flow. Teams can tweak spending as results roll in and markets shift. That freedom makes step-by-step upgrades and phased platform cleanup far easier. It also raises the bar for value recognition. Architects must continually show real progress to line up with the reforecasting.

Funding models will also impact the architecture approach. In a project setup, money dries up when the project closes. The system ships, gets tossed to operations, and improvement slows to a crawl. In contrast, a product model flips that script. Teams get steady funding to run, fix, and improve the platform over time. The payoff is tighter architectural discipline, steady refactoring, fewer surprise breakdowns, and real continuous improvement. It fits naturally with agile and DevOps ways of working.

Accounting rules steer tech choices in very real ways. Buy servers on-prem and you may treat them as capital expense. Pay for cloud and it often lands as operating expense. Same total cost, different accounting label. That label can sway executives even when the long-term math is close. Architects need to know these rules and sit down with finance to map the full impact. The best technical option can lose if it hurts short-term financial optics.

Cost Transparency and Technology Cost Allocation

Finance teams now expect clear visibility into technology spending. Showback and chargeback models assign costs to business units based on how much they use. Metrics such as cost per transaction, cost per customer, and cost per order have become standard management measures and they directly influence architectural choices.

Take cloud chargeback as an example. It depends on consistent tagging standards, clear resource categories, and reliable cost attribution. If infrastructure and applications aren't tagged consistently, finance can't assign costs with confidence. That means architects need to build financial traceability into the platform from the start. Policies for provisioning environments and consuming services aren't just operational rules; they act as financial controls.

Often, application rationalization starts with better cost visibility. When finance discovers several systems doing similar work, each with its own licenses and support contracts, consolidation quickly moves up the agenda. Enterprise Architecture supports this effort by linking business capabilities to applications and exposing overlap. When costs are clearly tied to systems, the argument for modernization or retirement becomes much stronger.

Cost visibility also shapes performance decisions. If leaders track cost per transaction, then scalability choices and automation efforts show up directly in financial results. Slow batch jobs or manual reconciliation processes translate into higher unit costs. Architects who understand these measures can design systems that not only work reliably but also run at a reasonable cost.

Financial Controls and Compliance Requirements

Financial governance frameworks such as SOX set clear requirements around access control, segregation of duties, auditability, and change management. These are not optional add-ons applied after system design. Segregation of duties, for example, means no single person can initiate, approve, and reconcile a financial transaction. Systems need to enforce role-based access controls that follow this principle. Identity management architecture therefore becomes closely tied to finance controls. Integration between HR systems and access provisioning platforms must be dependable and auditable.

Logging and traceability are just as important. Financial transactions need clear audit trails showing who did what and when. This shapes database design, application logging standards, and retention policies. Architects must make sure logs are tamper-resistant and available for audit review. Weak traceability creates compliance risk. It also erodes executive confidence.

Change management ties directly to the accuracy of financial reports. When you tweak an ERP setup or adjust the logic that decides how revenue is recorded, those changes can't be informal. They need to move through a clear, controlled path. The system should track versions and require formal sign-offs, with a built-in way to undo mistakes if something goes wrong.

External auditors, especially in heavily regulated fields, will dig into these controls and test whether they really work. If you bolt them on after the fact, expect higher costs and painful fixes. Building solid oversight into the architecture from the start saves time, money, and credibility down the road.

Financial Reporting and Data Dependencies

Financial reporting shapes many core data architecture choices. Statutory reporting demands compliance with accounting standards, a consistent chart of accounts, and dependable consolidation across legal entities. Management reporting, by contrast, typically calls for faster and more detailed insight. Executives now expect close to real-time visibility into financial performance, not just month-end summaries.

Meeting these demands puts ERP design squarely in focus. The selected platform, the way ledgers are configured, and how operational systems connect to finance all influence how quickly and accurately reports can be produced. When sales, inventory, and procurement systems rely on fragile batch integrations, reporting inevitably falls behind. Many organizations are moving toward event-driven integrations and standardized data contracts to improve both speed and accuracy.

Master data governance becomes a finance issue when inconsistent customer, product, or vendor records distort reporting. Duplicate records can inflate revenue or misstate expenses. Enterprise Architecture must define clear, authoritative data sources and governance processes. Finance leaders care deeply about data quality when it affects reported earnings.

Analytics architecture also hinges on financial reporting needs. Management may want profitability dashboards by product line or region. That demand shapes requirements for data warehousing, transformation rules, and consistent definitions. Architects should involve finance early to understand reporting hierarchies and key metrics. Designing data models without that input leads to rework and mistrust in the numbers.

Profitability Analysis and Unit Economics

Figuring out profit by product, customer, or sales channel takes more than pulling numbers from the general ledger. You need your day-to-day systems talking to your finance tools. Sales figures, cost behavior, service usage data, and cost allocation rules all have to line up cleanly. If they don't share the same logic and labels, the analysis falls apart.

Picture a company trying to see which customer segments actually make money. Invoice systems make revenue easy to track. But what about support time, shipping and fulfillment, marketing campaigns, and account management effort? Those costs live in different places. To assign them correctly, data has to flow across CRM, supply chain, operations, and finance systems. The setup must use shared IDs and common definitions so everyone counts things the same way, otherwise reports won't match, and trust in those reports erodes.

Unit economics bring the same challenge at a smaller scale. Leaders may ask for the true cost per transaction. Sounds simple, right? The system has to record transaction counts, cloud or server usage, third-party fees, and labor hours. Automated workflows should capture resource use as it happens and connect it back to the accounting system. When that link is missing, cost per unit becomes a guess, and less than trustworthy.

Profitability insights like these often force a reconsideration of strategy. A product that looks impressive from a revenue standpoint can turn out to lose money once every cost is properly assigned. Enterprise Architecture makes that clarity possible by keeping data flows complete, consistent, and traceable. Without that structure, the numbers can't be trusted. In the end, reliable financial insight rests on disciplined architecture.

Mergers, Acquisitions, and Financial Due Diligence

When a growth strategy includes mergers and acquisitions, finance usually leads the due diligence process. Financial statements, liabilities, revenue recognition practices, and cost structures are looked at closely. Yet behind every financial statement are systems that generate the data. Enterprise Architecture needs to quickly gauge how difficult integration will be.

Post-merger integration raises urgent architecture questions. Will the acquiring company combine ERP systems right away, or run parallel systems for a time? How will master data be aligned? What temporary reporting processes are needed to produce consolidated financial statements during integration?

Speed can make or break the value of a merger. If you push consolidation quickly, you may unlock synergies fast but move too fast and you risk breaking what already works. Keeping systems separate for a while can lower operational risk and give teams breathing room. Architects need to sit down with finance and weigh the real trade-offs: how fast to integrate, how much it will cost, and how much disruption the business can tolerate. Temporary setups, like interim data warehouses or mapping tables, often act as bridges before everything is folded into a single environment. They're rarely pretty, but they buy time and clarity.

Financial goals shape whether newly acquired platforms stay or go. When a deal is about gaining new capabilities, holding onto certain systems may be the whole point. When the goal is cutting costs, shutting platforms down quickly can be important. Architecture turns that financial intent into a clear, step-by-step path the teams can actually follow. Without it, how would anyone know what to retire, what to keep, and when?

Key Takeaways

Finance directly shapes Enterprise Architecture in practical, measurable ways. A company's financial strategy sets the tone for investment appetite, risk tolerance, and the speed of modernization. How budgets are structured, and how costs are accounted for, often determines how transformation programs are framed and funded. Demands for cost transparency also ripple into tagging standards, application rationalization efforts, and even performance design.

Regulatory and reporting requirements build controls straight into system architecture, influencing access management, logging, and change governance from day one. When leaders want clear profitability insights and tight unit economics, systems must support integrated data flows and disciplined master data management. During acquisitions, financial priorities often call the shots, shaping the order of integration and which systems stay, merge, or disappear.

CHAPTER 11

Information Technology

Information Technology holds a strange spot inside Business Architecture. It operates as its own business function while also powering almost every major initiative across the company. Marketing runs on digital tools. Operations lean on automation and clear, reliable data. Finance expects reports that are right and on time. Human Resources depends on integrated systems to manage talent. IT sits underneath all of it.

IT can't chart its course alone. The function has to convert company ambition into smart technology decisions, clear investment priorities, and practical guardrails that shape how systems are designed and run; choices that ripple outward for years. What happens when it doesn't? Technology turns into a patchwork of one-off projects instead of a unified set of capabilities.

The Kettera EA Methodology treats Information Technology as a strategic concern. Rather than seeing IT as a support desk in the basement, the methodology places it squarely in line with enterprise goals. That connection must be obvious, well-argued, and measurable, not implied. When alignment is fuzzy, focus drifts and systems multiply without purpose. When alignment is tight, technology becomes a coordinated engine for growth instead of a collection of disconnected tools.

IT Strategy Alignment with Enterprise Priorities

The first job of IT is to translate. When the business sets goals such as entering new markets, keeping more customers or cutting operating costs, those ambitions have to be turned into technology terms. Which platforms will support this shift? What capabilities are too weak? Where should spending speed up, and where should it pause?

Traceability keeps strategy from drifting off course. If leadership names digital customer experience as a top priority, the IT plan should point to clear, measurable actions: upgrading customer identity management, rolling out a unified customer data platform, building APIs for omnichannel engagement. Each effort needs a direct link to a specific business goal. When budget debates begin, someone should be able to ask, "Which objective does this fund?" and get a clear, specific answer.

Then come architectural guardrails. These are general boundaries, not rigid rules, that guide everyday choices. Suppose the company adopts a cloud-first directive; the IT strategy might define standards for infrastructure provisioning, identity management, and approved integration patterns. Teams still have room to adapt, but they move within a shared frame. That balance keeps work aligned without slowing it down.

When alignment is missing, companies accumulate duplicate systems, scattered data, and rising maintenance bills. Clear links between business goals and technology choices prevent that sprawl and tie every major investment to a concrete outcome.

IT Operating Model and Governance

How IT is structured shapes how architecture decisions get made. A centralized IT model typically applies standards consistently and brings specialized expertise together. This can reduce duplication and improve security oversight, but it may also slow response to business unit needs. In contrast, a federated or product-aligned structure puts technology teams closer to business domains, increasing agility but also raising the risk of fragmentation.

Enterprise Architecture must work effectively within whatever model is in place. In a centralized structure, EA may concentrate on enterprise-wide standards, shared services, and portfolio rationalization. In a federated environment, EA often serves as a coordination layer, making sure systems work together, data definitions stay consistent, and core principles are followed while still allowing domain-specific variation.

Governance is where the operating model shows up in practice. Who approves major technology investments? How are exceptions to standards managed? What triggers an architecture review? If budgets are allocated strictly by business unit, enterprise-wide platforms may struggle to find sponsorship. If funding is centralized, innovation at the edge may be limited.

Good governance gives teams room to move while keeping everyone pointed the same direction. Let people own their work, but tie it back to a shared plan. Architecture review boards and clear decision records make it work. They capture why choices were made and keep future work from drifting off course.

Digital Transformation and Modernization Strategy

Modernization is often framed as a technology refresh, but it is really a choice about business capability. When an organization moves to the cloud, consolidates ERP systems, or retires legacy platforms, the real question is what new abilities those changes unlock; quicker product releases, real-time insights, or stronger operational stability.

Take a retail company migrating from on-premise infrastructure to a cloud-based commerce platform. The benefit goes beyond reducing hardware expenses. What matters is the ability to scale instantly during peak seasons, use built-in analytics to personalize customer experiences, and roll out new features faster. Enterprise Architecture keeps these efforts connected instead of fragmented. Data migration, identity management, integration approaches, and security policies need to move in step across every program.

Platform consolidation offers another example. Many enterprises accumulate overlapping tools over time such as multiple CRM systems, redundant reporting platforms, and fragmented collaboration suites. Rationalizing these platforms cuts costs, but more importantly, it simplifies the operating model.

Training gets easier. Data flows more smoothly and innovation speeds up because teams build on shared foundations.

EA acts as the coordination layer across these efforts. It sequences modernization initiatives, identifies dependencies, and clarifies target states. Without that coordination, transformation programs compete for resources and add unintended complexity.

Enterprise AI Strategy and Governance

Artificial intelligence projects are popping up everywhere: factories use it to predict equipment failures, retailers tailor product suggestions, banks spot fraud, insurers speed up claims reviews. The activity is impressive but direction is not guaranteed. When AI efforts aren't tied to a clear strategy, they become science experiments with a budget.

What, exactly, should change because AI exists in your company? Maybe you want fewer customers leaving. Maybe you need sharper demand forecasts. Perhaps you hope to automate document-heavy workflows or speed up claims decisions. Every proposed use case must connect to a number that matters such as revenue gained, costs cut or time saved. Once that link is clear, technical direction becomes clearer: which data sources are approved, where models can run, how they plug into existing systems, and how their performance gets tracked over time.

Models can drift, misfire, or introduce bias, none of which you want showing up in tomorrow's headlines. Strong oversight means formal validation, ongoing bias checks, live performance tracking, and clear retraining timelines. Data needs scrutiny too. Is it complete? Is it consistent across systems? Was it gathered responsibly, with proper consent? Are there gaps that could skew results?

Enterprise Architecture brings order by defining common platforms and shared controls for AI work. Without that coordination, every department buys its own tools and builds its own models so costs climb and visibility drops. EA can create shared spaces for developing, testing, and deploying models, which lowers spending and makes oversight realistic. It also makes sure AI tools connect smoothly with core data and business applications, instead of becoming isolated tech islands.

AI-Driven Process Optimization and Automation

AI is no longer confined to analytics; it is steadily transforming day-to-day operations. By pairing machine learning with workflow platforms, intelligent automation takes over repetitive work such as invoice processing, claims adjudication, and customer service triage. At the same time, generative AI copilots support developers as they write code, help analysts draft reports, and assist service teams in responding to customer questions.

As these tools take on routine tasks, operating models and workforce structures begin to shift. Roles once centered on manual processing move toward oversight, handling exceptions, and higher-level analysis. Enterprise Architecture needs to plan for this transition. Can existing systems expose APIs that let automation tools connect smoothly? Do data pipelines enable near–real-time decisions? And is the compute environment able to scale for both model training and inference workloads?

Cybersecurity and Risk Posture as a Business Enabler

Cybersecurity usually gets labeled as defense but that view is incomplete. Protection matters, but security also fuels growth. People won't sign up for digital products they don't trust. Regulators won't greenlight expansion without solid safeguards in place. Partners hesitate to connect their systems unless they feel confident about yours.

A company's risk stance starts with a few clear ideas: zero-trust design, identity-based access control, strong encryption rules, and nonstop monitoring. These ideas guide how systems are built from day one. If identity becomes the main line of defense, then every application must tie into a shared identity service and follow the same login and verification patterns. Architecture choices flow from that decision.

Security teams and enterprise architects need to plan side by side. Separate them, and tension will build since projects slow down when security raises red flags at the last minute. Bring them together early, and security needs shape the design from the start, instead of at the end. The result is smoother delivery and fewer surprises.

Technology Talent Strategy and Capability Development

Technology strategy is limited by the skills an organization actually has. Great platform decisions mean little if no one can implement or support them. For that reason, talent planning and architecture decisions move together.

Some companies pour resources into in-house engineering, developing strong capabilities in cloud engineering, data science, and cybersecurity. Others depend more heavily on system integrators or managed service providers, while many land somewhere in between. That mix has real architectural implications. A heavily customized platform might be manageable for a large internal team, yet it can become a liability for a smaller organization that relies on outside partners.

The rise of AI-assisted development adds another layer. When developers work with AI copilots, productivity assumptions shift and training needs to evolve. Governance models also need to account for code quality and intellectual property risks. Enterprise Architecture has to reflect these factors when evaluating platforms and setting long-term direction.

Capability development is an ongoing effort. Certification programs, communities of practice, and structured knowledge sharing all play a role. Architecture is sustained not just by diagrams and standards, but by people who understand them and put them to use.

Innovation Management and Emerging Technology Evaluation

New tools show up frequently, each promising to change everything: blockchain-driven supply chains, IoT systems that predict equipment failures before they happen. Some of these bets pay off. Others quietly drain money and deliver little that moves the business forward. Who decides which is which?

Companies need a clear, repeatable way to judge what's worth trying. Pilot projects should have a firm end date, specific measures of success, and technical guardrails from the start. Enterprise Architecture plays referee here, checking that new trials fit current standards and don't introduce hidden links or fragile workarounds. Experiments shouldn't turn into permanent side systems by accident.

Too many rules can inhibit new ideas and innovation. The aim isn't control for its own sake; it's smart testing. EA can carve out "innovation sandboxes," contained spaces where teams explore new tools without putting core operations at risk. This provides freedom inside a fence. When a pilot proves its value, it earns the right to move forward, passing through formal review and careful integration planning before anyone talks about scaling it across the enterprise.

This balance keeps technology ambitions from pulling the strategy off course. The company stays open to new capabilities while staying grounded in its core intent.

Key Takeaways

Information Technology within Business Architecture is not just operational support; it serves as a strategic translator of enterprise intent into technological direction. Alignment with corporate objectives needs to be explicit, traceable, and reinforced through architectural guardrails. Operating models and governance structures determine how consistently standards are applied and how well investments are coordinated.

Modernization and digital transformation efforts become more coherent when they are sequenced and orchestrated through Enterprise Architecture. AI initiatives require clear alignment, strong governance, and standardized platforms to deliver lasting value. Cybersecurity, when built into architectural principles, becomes a growth enabler rather than a constraint. Talent strategy and innovation management help ensure that architectural ambitions stay achievable and adaptable.

With the Business Architecture domains: strategy, marketing, operations, finance, and IT, now defined, the focus shifts. Technology direction is aligned to business intent. The next step is to examine how that direction takes shape in concrete technical structures. We now turn to Technical Architecture, starting with the Application domain.

Technical Architecture

CHAPTER 12

Application

A company's goals might be built and a capability map might outline how the business is supposed to create value. None of that matters much until it shows up in the software people actually use. Applications are what employees sign into every morning and what customers click through every day. If Enterprise Architecture is supposed to shape how a company runs and changes over time, then applications are where those choices become concrete.

The Kettera EA Methodology places the application domain directly between business intent and technology delivery. Applications turn abstract capabilities into working features people can use. They are the bridge from idea to execution.

This chapter looks at how applications are categorized, designed, modernized, and eventually retired. Governance is woven through all of it. The goal is not to produce more or prettier diagrams. It is to build a clear, practical way to manage the enterprise application portfolio so it stays useful, current, and aligned with the business.

Application Portfolio Strategy

Most enterprises don't struggle with too few applications. They struggle with too many; introducing challenges such as overlapping systems that solve the same problem, aging platforms that drain time and money, and new tools introduced by individual departments. Without a clear portfolio strategy, that sprawl only gets worse over time.

Start by identifying what you have. Group applications into three broad categories: core, supporting or strategic. Core systems run the business and includes systems like order management for a retailer, claims processing for an insurer, patient records for a hospital. Supporting tools cover essential but non-differentiating work like payroll, email, or expense reporting. Strategic applications, by contrast, help the organization try something new: launch a business model, enter a new market, or significantly improve the customer experience.

These categories shape how you spend money and attention. Core systems need to be steady and dependable, yet modern enough to avoid becoming anchors. Supporting tools should be standardized and low-cost wherever possible. Strategic applications can justify more risk and quicker release cycles because they directly affect growth.

After categorizing comes rationalization. Review each application against a small set of criteria: business value, technical condition, maintenance cost, risk, and fit with the target architecture. Even a basic heat map can surface awkward realities such as two CRM systems serving similar users, a custom reporting tool that few people touch but everyone pays to maintain, or a legacy billing platform with rising support costs. Rationalization isn't a one-off cleanup effort; it's an ongoing habit built into the architecture review cycle.

Build vs Buy vs Partner

Every new capability raises a familiar question: should we build it ourselves, buy a commercial product, or work with a strategic partner? The answer should never rest only on developer enthusiasm or vendor marketing. It needs to tie back to differentiation and competitive advantage.

If the capability directly sets the organization apart in the market in ways like its pricing algorithms, proprietary logistics engine, or unique customer engagement platform, building may make sense. Custom development allows close alignment with business processes and ownership of intellectual property. But building brings long-term responsibilities: maintenance, security patching, upgrades, and retaining talent. A custom solution is not a one-time expense. It is a permanent commitment.

For non-differentiating capabilities, commercial off-the-shelf (COTS) or SaaS solutions are often a better fit. Few organizations gain an advantage by writing their own payroll system or collaboration tool. Buying lets the enterprise benefit from vendor scale and shared best practices. The architectural question then shifts to integration and data ownership rather than control of the code.

Partnership sits in the middle. In some cases, the organization may co-develop or embed within a broader ecosystem, for instance, a logistics provider integrating deeply with a major retailer's systems. In these situations, governance and contract structure matter as much as the technology itself. Who owns the data? Who sets the roadmap? Enterprise Architecture must guide these choices so partnerships don't create hidden lock-in or a fragmented architecture.

Build what makes you different. Buy what keeps the lights on. Partner or collaborate when scale or ecosystem reach really matters. License technologies when you need fast access without a long implementation.

Application Lifecycle Management

Every application moves through stages, whether we plan for them or not. It starts small, becomes more complex, levels out, and eventually loses relevance. When organizations pretend this cycle does not exist, technical debt builds up and risk steadily increases.

At the introduction stage, architecture reviews check that the system fits established standards, integration approaches, and security expectations. Budgets need to cover more than build costs or license fees; they should include ongoing operations and future improvements as well. Many initiatives move forward without a clear answer to a basic question: who owns the system after the implementation team walks away?

Over time, attention turns to refinement. Teams tune performance, improve usability, and add features in small increments to keep pace with business needs. This is also when technical debt becomes harder to ignore. Delayed upgrades, outdated libraries, and temporary fixes pile up quietly in the background. Enterprise Architecture should set clear triggers for action, for instance, when vendor support expires, when maintenance spending rises beyond a set share of replacement cost, or when release cycles slow to a crawl.

Retirement, however, is often avoided. Systems hang on because "someone might still need it" or because pulling out the data feels overwhelming. Defining objective criteria makes the decision easier: declining usage, full replacement of capabilities, or risk outweighing business benefit should all prompt planning. Decommissioning involves archiving data, ending contracts, and notifying stakeholders. A structured exit keeps outdated systems from lingering long after their value is gone.

Application Ownership and Governance

Applications fail as often from unclear ownership as from technical flaws. Every application should have both a business owner and a technical owner. The business owner is responsible for delivering value, setting enhancement priorities, and aligning funding. The technical owner makes sure the system follows architectural standards, runs reliably, and is managed through its full lifecycle.

Funding models reinforce accountability. Core systems may be funded as shared enterprise services, while strategic applications are often tied to specific business initiatives. When funding is unclear, responsibility is unclear too. Clear ownership also makes decision-making easier during incidents or audits because there's no confusion about who speaks for the system.

Architectural oversight does not mean central control over every design decision. It sets guardrails. Architecture review boards, lightweight design checkpoints, and clear standards create consistency while still letting teams move quickly. Governance should be proportional; a small enhancement to a supporting system does not need the same level of review as a new strategic platform that handles customer data across regions.

Architecture Patterns and Design Standards

If you use a similar playbook across apps life will get easier. Teams don't need one giant tech stack or a stiff template that fits no one. They do need shared patterns everyone understands and agrees to follow.

A company might settle on a modular monolith for mid-sized internal tools. It could choose microservices for large, customer-facing systems that must scale fast. Event-driven designs may fit work that cuts across multiple business areas. The pattern should match the job at hand, the experience of the team, and the reality of day-to-day operations. Microservices sound modern and powerful but they can also scatter your system across the network. Without strong DevOps habits and solid monitoring, they can cause more fires than they prevent.

Structure is only part of the story. Teams should align on how logs are written, how users sign in, how APIs are shaped, and how errors are reported. Shared rules make connections between systems far more predictable. New hires get up to speed faster because nothing feels random. But standards can't be frozen in time. Review them regularly or they turn into artifacts with no practical value.

Scalability, Resilience, and Performance Expectations

Availability and performance needs vary widely from one application to another. A customer-facing e-commerce site operates under far tighter expectations than an internal time-tracking system. Enterprise Architecture should sort systems into business criticality tiers, with each tier tied to clear non-functional requirements.

A Tier 1 application, for instance, may demand 99.9% uptime or more, disaster recovery across multiple regions, defined recovery time objectives (RTO), and ongoing performance monitoring. Tier 2 systems can often tolerate planned downtime and longer recovery periods. When technical standards reflect actual business impact, organizations avoid building elaborate solutions where they aren't needed or leaving key systems exposed.

Scalability also needs to be planned from the start. Expected growth in users, transaction volume, and data size should shape architectural decisions early on. Elastic infrastructure, thoughtful capacity planning, and thorough load testing belong in the design phase, not as last-minute fixes. It's much cheaper to plan for scale than to scramble for it later. With clear performance targets in place and metrics to track them, discussions move away from opinions and toward measurable service levels.

Application Rationalization and Redundancy Reduction

Redundancy often creeps in quietly. A marketing team signs up for a SaaS analytics tool while IT runs a separate reporting platform. Regional offices roll out their own ticketing systems. Shadow IT fills perceived gaps but ends up fragmenting the enterprise.

Application rationalization addresses this fragmentation. Through inventory analysis and capability mapping, overlapping tools come to light. The real question is which system best fits the future architecture, budget, and user needs? Consolidation may mean moving users, integrating data, or negotiating company-wide contracts.

Reducing redundancy cuts licensing costs, simplifies integration, and improves security oversight. It also makes accountability clearer. When one system is named the system of record for a given capability, reporting and data governance improve. Rationalization can be politically sensitive, as no team enjoys giving up “their” tool, but Enterprise Architecture brings the enterprise-wide view needed to make those decisions.

Bringing It Together

The application domain is where architecture planning creates user-facing results. Portfolio strategy keeps spending tied to what the company actually cares about. Choices about building or buying guard time, money, and attention. Lifecycle management keeps systems from going stale. Clear ownership and governance make sure someone is accountable. Shared patterns and standards create consistency without crushing initiative. Non-functional requirements connect speed, reliability, and scale to real business results. Rationalization cuts waste and trims unnecessary complexity, which most companies accumulate over time.

Put these pieces together and the application landscape stops being a random collection of tools and becomes deliberate. Every system has a purpose. Every system has an owner. Its role in the larger architecture is understood. Its future is planned, whether that means improvement or retirement. Clarity like this leads to better calls, faster adjustments, and the ability to change without chaos.

Next we'll go a layer deeper. Below the application layer sits data. Applications handle transactions and shape user experiences, but data lasts far longer than any single platform. Systems come and go but data remains. Knowing how information is organized, managed, protected, and shared across the enterprise is the next move in building a technology foundation that actually holds together.

CHAPTER 13

Data

Each order entered, insurance claim filed, shipment pushed back, new hire added, and bill settled leaves behind a mark. Stack up enough of those marks and you get something far more powerful than any single transaction: the company's memory; the hard-earned record of what really happened. Treated with care, that record becomes solid ground for smart choices and useful automation. Ignore it, and the same information multiplies, conflicts, and creates risk. What happens when no one agrees on which version of the truth is right?

In the Kettera Enterprise Architecture Methodology, data stands as its own architectural domain, with defined structure, ownership, controls, and a clear lifecycle. That means technical architecture must spell out how data is governed, organized, shared, protected, and eventually retired. When those rules are vague, or missing entirely, even thoughtfully designed applications can generate conflicting or unreliable results.

Data Governance and Ownership

The first question in any data discussion is straightforward: who owns it?

Many organizations automatically treat IT as the default owner of data because IT runs the systems. That assumption leads to confusion. Systems are technical assets but data is a business asset. A sales leader owns customer definitions. Finance owns revenue recognition data. Human Resources owns employee records. IT owns the platforms and databases that store and move that data but not the meaning or policies behind it.

A clear ownership model draws a line between **business data owners** and **technical custodians**. Business owners decide what the data means, who can access it, how long it should be kept, and what counts as acceptable quality. Technical teams put controls, backups, access management, and performance tuning in place. Between them sit **data stewards**, the people responsible for day-to-day data quality, resolving issues, and keeping definitions aligned across teams.

Formal governance structures support this model. A data governance council, usually led by a senior business executive with IT involved, sets policies and settles cross-domain conflicts. For example, if Marketing defines "active customer" differently from Finance, the council steps in and records a single company-wide definition. Without this forum, inconsistencies grow. Soon, reports contradict each other.

Governance isn't red tape. It's how a company keeps itself honest and on track. Strip away the jargon and you'll see it's simply a clear system for who answers for what, and who fixes problems when they pop up.

When a regulator starts asking questions, or the executive dashboard shows numbers that don't match, what happens? You shouldn't be scrambling for an owner. The team should already know who owns the issue, who makes the call, and who closes the loop.

Enterprise Data Model and Domain Boundaries

After ownership is established, the organization needs to decide how its data is structured at a conceptual level. That's where the Enterprise Data Model comes in.

This model is not a detailed map of every table in every database. Instead, it outlines major data domains such as Customer, Product, Order, Vendor, Employee, Financial Transaction, and others, and draws clear lines between them. Each domain carries agreed-upon definitions and documented relationships. An Order, for instance, is tied to a Customer and one or more Products. A Vendor provides Products. These links may sound straightforward, and they usually are, but different systems often represent them in inconsistent ways.

When domains aren't clearly defined, duplication creeps in. One application records a "client," another a "customer," and a third an "account." The names vary slightly, masking deeper differences in meaning. The enterprise data model addresses this by setting canonical definitions that are shared, approved descriptions of key entities and attributes. "Customer," for example, might be defined as "a legally recognized entity that has entered into a commercial transaction with the organization." From that point on, every system is expected to align with that definition.

Domain boundaries also reduce extra coupling between systems. If the Product domain owns product attributes, other applications should reference that domain rather than keep separate copies. Over time, this discipline pays off. It lowers reconciliation work and makes integration simpler.

Master Data Management (MDM)

Customer, product, vendor, and employee records show up everywhere, CRM, ERP, marketing tools, HR systems. Each platform keeps its own copy. When every system edits its own version, the details drift apart and no one knows which record to trust. Master Data Management, or MDM, steps in to fix this by creating one reliable source for each core entity.

An MDM approach builds what's often called a **golden record** for every key entity. Information from different systems feeds into a central hub, where software looks for matches and blends duplicate entries into a single record. Clear rules decide which data wins when two systems disagree. Suppose the ERP lists one billing address and the marketing tool lists another. The company might choose the ERP address as the official one, but let the marketing system's preferred contact method stand. Those decisions are set in advance through survivorship rules that spell out which source takes priority. Without that structure it's difficult to know which version to believe?

Synchronization patterns differ from one organization to

another. In a hub-and-spoke setup, the MDM platform distributes mastered records to subscribing systems. In contrast, a registry model connects records across applications without bringing every attribute into a single physical store. Factors such as scale, performance demands, and regulatory limits usually determine which model makes sense.

MDM isn't just a technical project. As mentioned earlier, some aspects of data management are technology focused but the meaning and interpretation of the data belongs to the business. Teams have to align on shared definitions, agree on data standards, and assign clear stewardship responsibilities. When this work is done well, day-to-day operations run more smoothly, sales and finance reference the same customer record, and service agents no longer have to guess which version is right which saves a lot of time and frustration.

Data Quality and Observability

Even with governance and MDM in place, data degrades without steady attention. Data quality must be defined, measured, and monitored.

Quality is often described across dimensions: accuracy, completeness, consistency, timeliness, and validity. These translate into clear, measurable thresholds. An organization might require that 98 percent of customer records include a validated email address, or that daily sales data be available in the warehouse by 6:00 a.m. each business day.

Validation rules enforce standards at the point of entry. Dropdown lists reduce free-text errors. Format checks confirm identification numbers. Referential integrity constraints ensure that an order cannot exist without a valid customer ID. But prevention alone isn't enough. Ongoing monitoring with data observability, tracks anomalies such as sudden drops in record counts or unusual shifts in data distribution.

Problems will certainly come up. When they do, you need a clear fix-it plan. Data stewards dig into what really caused the issue and make sure it gets corrected so it doesn't linger in the system and have to be addressed again.

If there isn't a process that tracks an issue from discovery to resolution, it will show up again. Over time, steady tracking through metrics and dashboards makes everything visible. Data quality stops being a fire drill and starts being something you measure, manage, and improve like any other performance goal.

Data Lifecycle Management

Some information supports day-to-day operations, while other records are kept mainly to satisfy compliance requirements. For that reason, system architecture needs clear rules for how data progresses from creation to archiving and, ultimately, deletion. When data is first created, required fields and validation rules set the baseline for quality. While it is actively used, systems emphasize speed and uptime. Over time, older data is often moved to less expensive storage, where it remains available for audits. Retention schedules eventually call for deletion, unless a legal hold interrupts the standard timeline.

Regulated industries face strict retention requirements. Healthcare records often need to be preserved for many years.

Financial transaction histories must often be kept for audit and tax purposes. Failure to follow these rules can lead to legal and financial consequences.

Lifecycle management also deals with cost. High-performance storage is expensive. Keeping ten years of rarely accessed transactional data in premium systems is hard to justify. Tiered storage strategies, hot, warm, and cold tiers, match cost to actual usage patterns. Transitions between tiers should be predictable and automated, not handled in an ad hoc way.

Analytics and Decision Support Architecture

Operational platforms record what has happened; analytics platforms figure out what it means. Keep them separate. When reporting queries run against the same systems that process orders or payments, slowdowns are almost guaranteed, and no one wants a checkout screen to freeze because someone pulled a quarterly report. Dividing the two protects day-to-day performance.

Most companies rely on some mix of a data warehouse and a data lake. Transaction data moves through ETL or ELT pipelines and lands in models shaped specifically for analysis and reporting. At the same time, firms are collecting large amounts of less tidy information such as application logs, IoT sensor streams, chat transcripts and social media posts and parking it in data lakes built to handle that messiness. Not all data needs to be forced into a columnar format, as modern analytics tools can work with unstructured data.

If leadership measures customer acquisition cost, churn rate, average order value, and inventory turnover, the data setup has to produce those numbers the same way every time. A KPI shown in one dashboard should match the same KPI shown somewhere else. Every calculation needs to point back to a single, agreed-upon enterprise definition so teams aren't arguing about whose spreadsheet is right.

Self-service business intelligence tools let business users explore data on their own. But freedom without guardrails leads to inconsistent numbers and conflicting reports. Certified datasets, semantic layers, and governed metric definitions strike the right balance with room to explore and clear standards for official reporting.

Data Security and Privacy

Data architecture carries real risk. Sensitive information such as personally identifiable information (PII), financial records, health data must be handled carefully.

Classification standards group data based on sensitivity: public, internal, confidential, restricted. Each category ties to specific controls. Restricted data may require encryption at rest and in transit, multi-factor authentication for access, and detailed audit logging. Internal data might have fewer limits but still needs access control.

Privacy regulations such as GDPR and HIPAA add extra requirements. Individuals may have the right to request access to their data, ask for corrections, or require deletion under certain conditions. The architecture must support these rights. It's not enough to store data securely. Organizations must also know where it lives and how it moves across systems.

Managing risk takes more than ticking boxes on an audit form. Teams need steady access reviews, real penetration tests, clear incident response plans, and regular response drills to make those plans real. These are basic protections, not extras you add when there's time.

Security has to be built in from day one, when possible. You can't roll out a system and then try to glue safeguards onto it later without expecting some gaps. Design choices shape exposure, and once those choices are locked in, fixing them gets messy and expensive.

Data Integration Patterns

Data rarely lives in just one application. Orders travel from e-commerce sites to ERP systems, and customer updates pass from CRM into billing. How you move that data affects speed, consistency, and how hard the system is to manage.

Batch integration is still widely used for large volumes that don't need instant updates. Nightly file transfers or scheduled ETL jobs refresh downstream systems at set times, creating predictable processing windows. By contrast, real-time integration, through APIs or event-streaming platforms, keeps systems in sync as changes happen and supports fast user feedback. If a customer changes an address, for instance, other systems may need that update almost immediately.

Event-driven architectures use publish-and-subscribe (pub-sub) models, where systems emit events like "OrderCreated" or "PaymentReceived." Subscribers respond on their own, which reduces tight coupling. ETL and ELT approaches decide where transformation happens, before loading into the warehouse or after. Each option affects performance, scalability, and governance.

Data architecture needs to align with integration standards set in the broader technical architecture. Consistent patterns make maintenance easier and help prevent operational surprises.

Bringing It Together

Data architecture isn't just a collection of random rules. It's a cohesive system. Governance assigns ownership and responsibility. The enterprise data model gives everyone the same language. MDM keeps core entities synchronized. Quality checks preserve trust. Lifecycle rules rein in cost and meet compliance demands. Analytics tools turn raw data into decisions. Security locks down what can't be exposed. Integration patterns move information cleanly from one system to the next.

Get these pieces working together and people believe the numbers. Leaders stop second-guessing dashboards. Frontline teams act on system outputs without workarounds. Regulators get reports that hold up under scrutiny. Swap out an application, or overhaul an entire platform, and the company still remembers how to operate.

Next comes the connectivity. Integration architecture sits on top of the data groundwork described here and spells out how systems talk and how information travels across the business, reliably, every time. Ignore integration discipline and even beautifully governed data stays trapped in silos.

CHAPTER 14

Integration

Applications generally work fine on their own. Challenges arise in the gaps between them, where data passes from one system to another or where ownership gets fuzzy. Those weak spots decide whether software works together smoothly or pulls in different directions. Integration acts as the enterprise's connective tissue. It determines whether separate tools function as a unified whole or sit side by side with little coordination.

Inside Technical Architecture, integration defines how applications and information travel across business areas. The sales platform leans on finance to close the books. Operations tools pull from shared customer records. Partners outside the company need tightly managed entry points into specific features.

Organizations without a deliberate integration approach end up stitching systems together with fragile, one-off links and hidden dependencies that make every change risky. The result is software that feels stuck and even small updates ripple outward in unpredictable ways. A thoughtful strategy flips that script. Integration becomes an engine for growth and flexibility, not a bottleneck that slows everything down.

Enterprise Integration Strategy

Every integration strategy starts with a point of view. Is the company API-first? Event-driven? Built around batch processing? Most large organizations blend these approaches, but where they put the weight makes a real difference. The strategy should match how the business actually operates and plans to grow, not whatever approach happens to be popular at the moment.

In an API-first model, business capabilities are exposed as clearly defined services. Customer profile lookup, order submission, pricing calculation, each function is available through documented interfaces. This setup fits companies that value modular design, reuse across teams, and easy connections with partners. Consider a retailer moving into online marketplaces: clean, public-facing APIs let third parties plug in without custom integrations which saves everyone time and headaches.

Event-driven architecture works best when responsiveness and decoupling matter most. Instead of one system calling another directly, systems publish events such as “OrderCreated” or “PaymentReceived,” and subscribers respond asynchronously. This approach fits high-volume, real-time operations required for fraud detection, inventory updates and logistics tracking, where state changes need to spread quickly without tightly linking systems.

Batch-oriented integration still has its place. Financial

reconciliation, regulatory reporting, and large data migrations often run on scheduled data transfers. Trying to turn every nightly job into a real-time service can add unnecessary complexity. A strong strategy recognizes where delay is acceptable and where it isn't.

Most enterprises end up with a hybrid model. The architecture team should clearly document when to use synchronous APIs, when to publish events, and when scheduled data exchange makes sense. These choices need to connect directly to business capabilities. A capability that requires immediate user feedback, such as credit approval, likely calls for synchronous interaction. Analytics-focused capabilities can handle delayed data ingestion.

API Architecture and Management

APIs help define the shape of integrations. They spell out how systems talk to each other and how outside developers tap into company capabilities. Technical Architecture must lay down clear design standards and stick to them. Most transactional services lean on REST with HTTP and JSON because it's simple and widely understood. GraphQL fits better when users need flexible queries, especially inside complex interfaces. Select the approach based on purpose, not by habit.

Versioning demands discipline and clear rules. Teams should assume backward compatibility unless there is a strong reason to break it. If a breaking change can't be avoided, the new version must be clearly labeled and the timeline shared early. Publishing a deprecation schedule ahead of time helps teams plan and prevents last-minute surprises.

API gateways act as a central hub for routing, enforcing authentication, applying throttling rules, and collecting analytics. They give both internal and external users a single, consistent entry point. With rate limiting in place, backend systems are shielded from traffic spikes, and usage data helps teams plan capacity with more accuracy. When offering public APIs, organizations benefit from a developer portal that includes documentation, code samples, and self-service key management, which can significantly cut down support requests.

Internal APIs deserve as much rigor as external ones. Too often, teams treat internal interfaces casually, assuming informal coordination is enough. But internal consumers can be just as dependent on stability. Clear ownership, solid documentation, and lifecycle management matter just as much inside the organization.

Event-Driven Architecture

APIs create direct connections but event-driven systems handle tasks in a more indirect manner. Instead of one system asking another for data, systems broadcast events over messaging or streaming tools, and other systems react when they're ready. Because producers and consumers don't depend on each other directly, you avoid tight coupling and make it easier to grow.

At the core is the publish–subscribe model. One system sends out an event and doesn't care who's listening. Interested subscribers sign up and act when the event appears. The publisher doesn't need to change when a new consumer shows up. Picture an "OrderShipped" event. Shipping trackers update. Customers get notifications and analytics dashboards log the data. All separately and at their own pace.

Event structure matters. A small tweak to an event's format can break dozens of downstream systems. That's why teams use schema registries and clear compatibility rules to control changes. Perfect alignment across systems isn't guaranteed, either. In event-driven setups, different services may see slightly different data at the same time. Leaders need to understand that trade-off, speed and flexibility in exchange for brief mismatches. For things like inventory counts or loyalty points, a short delay rarely hurts. Financial postings might be another story.

Real-time needs often make it worth some additional complexity. Fraud detection tools scan transaction streams as they happen. Operations dashboards refresh nonstop. But should every system be built this way? Of course not. Running an event-driven platform means watching for slow consumers, managing replays when something fails, and keeping the plumbing healthy.

Integration Patterns and Standards

As organizations grow, their integration patterns tend to change. In the beginning, many teams wire systems together with simple point-to-point connections. They're fast to set up, which makes them attractive early on, but scaling them is another story. With every new link, dependencies multiply, and even small changes can impact systems in unpredictable ways.

To bring order to the sprawl, some companies adopt a hub-and-spoke model built around a central broker or integration platform. Instead of connecting directly to one another, systems plug into the hub. Routing and data transformation become easier to manage, though a poorly designed hub can turn into a choke point. Traditional Enterprise Service Bus (ESB) architectures follow this approach, relying on centralized orchestration and mediation to coordinate traffic.

More recent designs lean toward microservices paired with lightweight messaging, trimming back heavy central coordination. Integration Platform as a Service (iPaaS) offerings provide cloud-based connectors, data transformations, and workflow automation, especially handy when tying together SaaS applications. ESB platforms concentrate control in one place, which can limit flexibility over time, while microservices give teams independence but demand disciplined governance to prevent sprawl.

The architecture team should document approved patterns and clarify when to use each one. For example, point-to-point integration may be allowed for temporary or low-risk connections. Long-term, strategic integrations should use standardized APIs or event-driven approaches. Clear guidance will prevent the systems from deviating from agreed upon patterns.

Integration Governance and Ownership

Most integration breakdowns trace back to unclear ownership. When no one clearly owns an API, data contracts get broken. When the responsibility for an event schema is unclear, updates can conflict. When shared transformation logic "belongs to everyone," it ends up maintained by no one. Every integration needs named owners and written accountability.

Each API needs a specific product owner who handles its full lifecycle and supports the people who use it. Data contracts that connect systems demand shared care, especially when they span different business areas. Integration platforms such as ESB, iPaaS, streaming tools or whatever you run, also require a dedicated operations team with clear service targets and uptime commitments. Without that structure, it becomes unclear who can make appropriate decisions.

Review processes protect the quality of systems. Architecture review boards should evaluate new integrations for security risks and long-term supportability. Version control is not optional and must cover APIs, event schemas, and other integration assets. Change management needs clear impact analysis and direct communication plans, particularly when outside partners depend on those connections.

Security and Access Control in Integration

Each integration point creates a possible exposure. Once APIs are live or events are moving across networks, it's too late to start thinking about security. To produce the best security models, they should be built in from the start, not treated as an afterthought.

Set authentication and authorization standards at the enterprise level rather than leaving teams to decide on their own. User-facing APIs typically rely on OAuth2 and OpenID Connect, while service-to-service calls often use mutual TLS with short-lived tokens. Clear token management policies should define how tokens expire, how they are rotated, and how they can be revoked. Storing hard-coded credentials in configuration files is poor practice.

When exposing internal services to external partners, you need extra controls. API gateways can enforce rate limits, validate tokens, and inspect payloads. Network segmentation and zero-trust principles help limit how far an attacker can move. Architects should also consider data minimization, only exposing the fields needed for the intended purpose.

Security testing must go beyond application code. Integration endpoints need penetration testing, schema checks, and monitoring for unusual patterns. Treat every boundary as a controlled interface, not a trust zone.

Integration Observability and Reliability

Operational visibility is everything after integrations go live. Distributed systems don't always crash in obvious ways; they fail quietly, through timeouts, half-delivered messages, or data that gets altered without anyone noticing. Without one place to see logs and trace activity across services, teams are left guessing during root cause analysis.

Monitoring has to cover what really matters: how fast APIs respond, how often they fail, how much traffic they handle, and whether the services they depend on are even healthy. Event-driven architectures add another layer of risk, so you need insight into message queues, consumer lag, dead-letter queues, and even retry backlogs. Tracing tools tie requests together across services, letting you follow a single transaction from start to finish. If a customer says a payment didn't go through, shouldn't your support team be able to track the activity related to that payment?

Retry rules and failure responses should follow shared

standards across teams. Automatic retries help with brief network hiccups, but done poorly they can flood struggling systems with even more traffic, making a bad outage worse. Circuit breakers act like safety switches, stopping failures from rippling outward and protecting everything downstream. Clear service-level agreements set expectations for uptime and response times, making ownership and accountability clear.

Operational dashboards built for both business and technical teams close the loop. Business leaders can watch order volumes in real time, while engineers keep an eye on latency and overall system health. Each view tells part of the story and you need both to see the full picture.

Legacy and Third-Party System Integration

Few enterprises are building on greenfield platforms. Legacy systems, including mainframes, proprietary databases, aging COTS applications, are still deeply embedded in core processes.

Adapters and anti-corruption layers create a buffer between modern services and legacy complexity. Instead of letting new applications depend directly on outdated schemas or protocols, a translation layer converts data into standardized formats. This shields the broader architecture from legacy constraints and makes future replacements easier.

Third-party SaaS providers bring additional challenges. Vendor APIs can change with little or no notice, rate limits can complicate designs, and data residency regulations may restrict integration patterns. iPaaS solutions can simplify connectivity, but they must be assessed for scalability and cost.

Working with outside partners means spelling out, in writing, how stable the interface must remain, how quickly the team will respond and escalate issues, and how much advance notice they owe you before making changes, avoiding surprises and vague promises.

Key Takeaways

Integration determines whether an enterprise behaves like a coordinated system or a collection of disconnected tools. A clear strategy such as API-first, event-driven, batch-based, or a mix, connects technical choices to business capabilities and plans for growth. Setting standards for API design, event schemas, and approved integration patterns cuts down on inconsistency and keeps the architecture from drifting over time.

Ownership and governance need to be explicit so no one is guessing who is responsible for what. Security has to be built into every integration boundary rather than added later. Strong observability and reliability practices help teams spot and fix problems before they grow. When working with legacy or third-party systems, careful insulation protects the rest of the architecture and saves you from surprises down the road.

With integration defined, we can now look at the underlying infrastructure that supports these capabilities. The next chapter explores Platform Architecture, including the compute, storage, network, and cloud environments that allow applications and integrations to run at enterprise scale.

CHAPTER 15

Platform

Applications and data must live somewhere. Every tool a company uses relies on core plumbing: networks, identity systems, automation pipelines, and monitoring tools. Strip those away and nothing runs. That "somewhere" is the platform, the ground everything else stands on. In enterprise architecture, designing a platform sets the pace for delivery, shapes risk and drives spending decisions.

Business Architecture defines the goals. Application and Data Architecture organize the apps and information. Platform Architecture answers the questions: Where does it run? How is it released? How is it locked down, watched, supported, and paid for?

A strong platform strategy brings consistency and defines clear boundaries while leaving room to flex. Striking that balance takes clarity across several areas: the cloud operating model, standards for compute and storage, networking, identity, DevOps, observability, resilience, and cost management. Each of these pieces fits together to form a coherent whole.

Platform Strategy and Cloud Operating Model

Every enterprise must make a clear decision about its cloud posture. Single cloud? Multi-cloud? Hybrid with significant on-premises investment? The answer shouldn't be driven by trends or vendor marketing. It needs to reflect regulatory constraints, existing investments, geographic footprint, vendor risk tolerance, and internal capabilities.

A single-cloud strategy simplifies governance and lowers operational complexity. Multi-cloud can reduce vendor lock-in and meet regional needs, but it adds integration overhead and splits expertise across platforms. Hybrid models are common for organizations with legacy systems or data residency limits. Choosing the appropriate model requires architectural intent and understanding.

Once the high-level strategy is set, the enterprise must design its landing zones. A landing zone isn't just a cloud account, it's a preconfigured environment with networking, identity integration, logging, security controls, and policy guardrails already in place. Standardized landing zones ensure new workloads begin in a compliant state instead of being fixed later.

Tenancy and subscription setups need to mirror the company's business domains as defined in Business Architecture. Give each major area, Marketing, Operations, Finance, even Product, its own subscription or account, with separate budgets and clear security lines. Money stays in its lane and access stays controlled.

Governance guardrails like required encryption, approved regions, strict tagging rules, and access controls get locked in through policy engines. That's Enterprise Architecture in action; turning big-picture principles into designs that actually hold up under pressure.

Compute, Storage, and Runtime Standards

Platform architecture needs to spell out which compute options are approved and when to use each one. Some legacy systems or specialized software still require virtual machines. Containers and Kubernetes are better suited for scalable, portable applications, while serverless functions handle event-driven workloads without much infrastructure overhead.

The standards must clearly explain where different workloads belong. For example, a customer-facing web app with unpredictable traffic could run on container orchestration with auto-scaling. Scheduled serverless environments may be a better fit for batch jobs. If a commercial off-the-shelf product comes with strict vendor requirements, managed virtual machines might be the safest choice. When this guidance is missing, teams tend to stick with familiar tools instead of selecting the best option for the job.

Storage needs the same level of clarity. Transactional databases often demand high-performance block storage, unstructured data fits well in object storage, and long-term records belong in archival tiers. Each option carries different cost and performance trade-offs. Lifecycle policies can automatically shift data between tiers based on how old it is and how often it's accessed, which saves teams from moving it by hand. By defining these policies at the platform level, organizations prevent every team from creating its own version.

Elasticity expectations must be clearly defined. Are workloads required to scale automatically? Must environments shut down outside business hours? Should development environments be short-lived and recreated from code? They shape cost, reliability, and the developer experience. The enterprise should document infrastructure lifecycle standards including provisioning, patching, decommissioning, and configuration management.

Network Architecture and Connectivity

In the cloud, everything starts with a Virtual Private Cloud (VPC) or Virtual Network (VNet). That's the basic container for your traffic. From there, you need clear boundaries: a shared services network for things like identity and logging, application-specific networks to isolate workloads, restricted zones for sensitive systems, and even a separate management network for admin access. These segments shouldn't talk to each other freely. All traffic should move through inspection points such as firewalls or secure gateways, that check and filter what passes between them.

Hybrid connectivity introduces additional complexity. Links between your data center and the cloud must be both secure and redundant. A simple VPN tunnel might handle lighter workloads. Heavier traffic often demands dedicated circuits such as Direct Connect or ExpressRoute, which offer steadier performance and lower delay. Name resolution is also important. Your DNS approach should work the same way everywhere so systems can reliably find each other. Split it into disconnected pieces and you invite mysterious outages that could take days to untangle.

Traffic management patterns, such as load balancing, API gateways, and content delivery networks, work best when they align with agreed-upon reference architectures. If each team designs its own ingress approach, complexity increases unnecessarily. Enterprise Architecture defines the sanctioned topologies and clearly records the situations where deviations are allowed.

Identity, Access, and Zero Trust Foundations

An enterprise IAM strategy should centralize authentication while letting teams grant access where needed. Federation with corporate directories lets users sign in once across cloud and on-premises systems. Role-based access control (RBAC) sets permissions by job function, while attribute-based access control (ABAC) adds context limits such as location or device posture.

Privileged access needs additional attention. Administrative rights shouldn't live forever, and every action taken with them must leave a digital trail. These powerful accounts should stay separate from everyday user logins to reduce mistakes from lack of attention. Emergency "break-glass" access should exist, but it must sit behind strict oversight. Tight controls like these shrink the damage a stolen credential can cause.

Zero trust doesn't stop at logging users in. Systems talking to other systems must prove who they are and confirm what they're allowed to do, every single time. Being on the right network means nothing by itself. Location alone doesn't equal trust. Data must stay encrypted while moving across the wire and while sitting in storage. The platform builds in these rules automatically, so app teams aren't forced to bolt them on later and risk doing it differently each time.

DevOps, Automation, and Platform Engineering

Modern platforms are defined as much by how they automate work as by the infrastructure they run on. Relying on manual provisioning slows teams down and invites mistakes. With Infrastructure as Code (IaC), environments can be version-controlled, reviewed by peers, and recreated reliably whenever needed.

Most enterprises put standardized CI/CD pipelines in place that bundle code scanning, security testing, artifact management, and automated deployment into a repeatable flow. While teams can build on top of these pipelines, the core stages stay the same across the organization. That consistency is important. Keeping those foundations uniform limits unnecessary variation and makes it easier for new teams to get started quickly.

Platform engineering pushes this idea further by building internal developer platforms (IDPs). With self-service portals, development teams can request environments, deploy services, and access monitoring dashboards without having to submit infrastructure tickets. Guardrails still exist, but there's less friction. This doesn't remove governance. It just builds governance directly into the automation.

Observability and Operational Management

A platform you can't see clearly is hard to run well. When possible, every environment should follow the same rules for logging, metrics, distributed tracing, and alerting.

Bring all logs into one shared system so security issues and day-to-day failures show up in a single place. When data lives in scattered tools, teams waste time hunting for answers, time you rarely have during an outage. Metrics need to track the basics that matter most: the health of your infrastructure and how your applications are performing. Without that, you won't know something is slowing down until users start complaining.

Service Level Objectives (SLOs) and Service Level Agreements (SLAs) need to match the real stakes of the business. A payment processing platform might demand 99.99% availability, while an internal reporting tool can live with looser targets. Dashboards should mirror those expectations, showing at a glance when performance slips past defined limits. Document incident response procedures such as escalation routes, roles, and post-incident reviews, and run through them regularly because no one wants to figure it out mid-outage.

Resilience, Backup, and Disaster Recovery

Not all systems need the same level of resilience. Platform architecture should define availability tiers based on business impact. Tier 1 systems might require multi-region deployment with automatic failover. Tier 2 systems may rely on single-region high availability with fast restore capabilities. Lower tiers can accept longer recovery times.

Recovery Time Objective (RTO) and Recovery Point Objective (RPO) targets should be set in partnership with business stakeholders. These targets guide technical decisions like replication frequency and backup intervals. Without clear targets, disaster recovery becomes guesswork.

Backups must run on their own, stay locked down with encryption, get monitored, and be tested. Yet plenty of teams only learn the hard way that the copies they trusted are partial, corrupted, or impossible to restore. What good is a backup if you can't restore it? Automating the process removes human errors, and strong encryption keeps the data private, especially when data lives offsite, where you don't fully control the hardware. Additional reliability comes from copying data across regions, using storage that can't be altered, scheduling regular failover drills, and reviewing who can access what. None of that feels urgent on a calm Tuesday afternoon but during an outage, it will become obvious if it's missing.

Cost Management and FinOps

Spending in the cloud is almost too easy. What's harder is seeing clearly who is spending the money and for what purpose. To manage that risk, organizations put a formal FinOps model in place. Clear tagging rules connect each resource to a business unit, project, and cost center, so expenses aren't left unclaimed. Showback reports give leaders visibility into how much they're using, while chargeback models assign the real costs to the teams responsible. Budget alerts, and even automatic shutdown policies, help stop runaway spending before it gets out of hand and before finance starts asking uncomfortable questions.

Optimization practices should be ongoing. Rightsizing instances, removing idle resources, and committing to reserved capacity when it makes sense are good operational habits. Platform teams need to provide clear visibility so business leaders can weigh cost against value. Architecture without financial accountability is incomplete and violates one of the core concepts of Enterprise Architecture, which is that it should always tie back to business results.

Key Takeaways

Platform Architecture is the physical bedrock of every enterprise system. It sets the cloud direction, locks in standards for compute and storage, shapes the network, pulls identity into one place, hardwires automation, builds in visibility, prepares for failure, and keeps spending in check. None of this stands alone. One weak spot such as security, cost control or resilience, and the impact can ripple through the entire company.

Applications, Data, Integration, and Platform have been defined but what do Enterprise Architects actually create day to day? The next section dives into the real outputs: the artifacts that turn big-picture thinking into concrete guidance teams can use and leaders can measure.

Enterprise Architecture Artifacts

CHAPTER 16

Defining What Enterprise Architects Do

Ask ten professionals to explain what an Enterprise Architect actually does and you'll hear ten versions of the job. One person sees a seasoned technologist critiquing solution designs. Another imagines a big-picture thinker who only meets with executives. Someone else assumes it's a former developer who has stepped away from coding but still hovers near it. No wonder people get confused. The role lives where business plans, technology delivery, and company oversight meet.

Titles don't always help either. One clearer way to define the role is to look at what Enterprise Architects actually produce. What do they deliver? How do they go about creating it? When you focus on outputs and repeatable ways of working, the role becomes clearer. In the Kettera EA Methodology, architecture isn't about vague influence or positional power; it's about concrete artifacts that steer real decisions across the company.

Architecture as a Producer of Artifacts

Enterprise Architects create artifacts that steer change. These are not academic papers and diagrams. They are structured views of how the organization works today, where it intends to go, and how it plans to get there. Capability maps, future-state diagrams, technology standards, roadmaps, and decision records all belong in this set.

What sets these artifacts apart is their reach. A solution architect might design for one project. A senior developer might shape the inner workings of a single application. Enterprise Architects operate on a wider canvas: their artifacts cut across systems, business units, and funding cycles.

Take a straightforward case like a development team that wants to adopt a new customer relationship management platform. The senior developer checks technical fit. The solution architect plans the integrations. The Enterprise Architect steps back and asks broader questions: Does this support our defined business capabilities? Are we duplicating tools we already own? Where does this sit in our 24-month roadmap? The resulting artifact, an updated portfolio view or a revised roadmap, affects not just this effort, but the direction of future spending.

Differentiating Architecture from Senior Development

The line between architecture and advanced engineering is

often blurred in real-world practice. Many Enterprise Architects started their careers as developers, and they bring that technical depth with them. But experience alone doesn't define the role. What Enterprise Architects do is fundamentally different from what senior developers and engineers do. Senior developers concentrate on building systems well. Enterprise Architects focus on building the right systems.

The gap sounds minor but it isn't. Developers tune code and performance inside a box someone else drew. Enterprise Architects draw the box. They decide which capabilities get funding, which tools are used and which systems can connect to others. A developer may argue about one framework versus another; the architect steps back and asks a bigger question: does either choice fit the company's long term strategy?

These teams operate in different lanes with different cycles. Dev teams move through familiar build cycles; plan, code, test, ship. Enterprise Architects operate on governance tracks and long-range planning. They run architecture reviews, guard the standards library, measure how mature the business's capabilities really are, and steer roadmaps that stretch across programs and years. Their job overlaps with engineering, but it doesn't stop there; it extends across the whole organization.

An experienced developer can shape the technical direction of a team or product through hands-on decisions and practical insight. An Enterprise Architect takes that direction and turns it into structured, organization-wide guidance that others can follow. They connect individual technical choices to the broader business strategy.

Differentiating Types of Architecture

Even within architecture itself, the boundaries can blur. What distinguishes Enterprise Architecture from Business Architecture, Solution Architecture, or Technical Architecture? The answer comes down to scope and deliverables.

Business Architecture artifacts explain how a company works and what it is trying to achieve. They lay out the abilities the organization needs and how value flows to customers. Technical Architecture artifacts spell out the ground rules for applications, data, integration, and platforms. Solution Architecture artifacts zoom in on one initiative and show how it will be built. Enterprise Architecture pulls these viewpoints together so they make sense as a whole.

An Enterprise Architect might draft a capability map that shows what the business must be able to do. A Technical Architect sets the approved integration patterns that make those abilities possible. A Solution Architect then uses those standards on a real project, shaping them to fit the situation at hand. The Enterprise Architect checks that all of this lines up. Enterprise Architecture makes sure the big picture and the day-to-day work actually align.

Enterprise Architecture acts like connective tissue, linking strategy to systems and plans to code. It doesn't take over the roles of other architects; it guides and coordinates them. Its purpose is simple but demanding: carry business intent through to execution so that every project follows a steady, consistent technical direction across the organization.

The Role Defined by Process

Artifacts by themselves don't define the role. Just as important is the way those artifacts are created, reviewed, and maintained. In the Kettera EA Methodology, architecture follows a steady cadence; a repeating cycle of revisiting strategy, refreshing current-state views, setting future-state targets, and adjusting roadmaps as conditions change.

At the center of this rhythm are structured conversations. Enterprise Architects sit down with business leaders to clarify goals and constraints. They work closely with technical teams to test what's realistic and what isn't. Before major investments move forward, they lead formal reviews to examine alignment and risk. Over time, this steady cadence turns architecture artifacts into living tools that guide decisions rather than documents that gather dust.

For example, a capability map should not be created once and archived. It needs to be reviewed when strategy shifts or when new regulatory requirements appear. A roadmap should change as projects wrap up, priorities shift, or funding changes. The Enterprise Architect owns that activity. That ownership, of both artifact and process, is what sets the role apart from a one-time consultant or an informal advisor.

Enterprise Architects as Decision Enablers

Enterprise Architecture lives or dies on the quality of decisions. Senior leaders constantly decide what to invest in, which systems to shut down, which standards to lock in, and what to focus on next. Architects don't own all those calls but they do shape how those calls are made.

Picture a leadership team weighing two rival proposals. Without architectural input, the loudest presenter or the rosiest short-term forecast might win the day. Bring in architectural views, and the tone will shift. Which option actually moves us closer to the future we agreed on? What new technical debt are we taking on and is it worth it? The architect may not dominate the room, but their groundwork guides the outcome.

Good architecture doesn't control decisions. It makes smart ones easier to determine.

The Governance Dimension

Architecture without governance becomes little more than advice. Enterprise Architects put review processes in place so standards are followed and strategy stays on track. This is not about standing at the gate and stopping progress. It is about setting up a clear, reliable way to evaluate proposals so every team plays by the same rules.

Typical governance artifacts include review templates, compliance checklists, and documented decisions. Far from being red tape, these tools create transparency. Teams understand ahead of time how their designs will be measured, and leaders can see that proposals have been checked against enterprise standards before any money is approved.

Suppose a project team wants to adopt a new database technology. Governance requires them to explain why the current approved technologies aren't enough, describe how the new tool would connect to existing systems, and review its impact on day-to-day operations. The Enterprise Architect reviews this submission, may consult with data architects, and then records the final decision. Over time, these decision records become a valuable institutional memory. They show what the standards are and why.

Strategy Translation

Translating business strategy into clear architectural action sits at the heart of the Enterprise Architect's role. Leadership teams talk in broader goals such as entering new markets, delighting customers, cutting costs and growing digital revenue, but those statements rarely tell delivery teams what to actually build. What does "improve customer experience" even mean to a system owner staring at a backlog?

Enterprise Architects take those big ambitions and recast them as concrete structural choices. A push into new markets could call for stronger distribution networks, systems that handle multiple languages, updated compliance capabilities, and scalable infrastructure that can handle sudden growth. A mandate to improve customer experience might require unified data across channels and tighter integration between front- and back-office platforms. Cost reduction often points to trimming duplicate applications, simplifying the technology stack, modernizing aging systems, or renegotiating vendor contracts.

Architecture turns strategy into wiring, platforms, data flows, and standards that teams can actually execute against. The artifacts make strategy real. A future-state diagram can map out a unified customer data platform, while a two-year roadmap lays out the order of consolidation efforts. Standards may tilt toward cloud-native technologies to handle growth. In this way, big-picture intent turns into clear, practical direction.

Where Architecture Ends

It is also useful to define what Enterprise Architects do not do. They do not manage projects. They do not write production code, at least not as part of their core responsibilities. They do not replace business leaders in setting strategy. Their authority is architectural, not day-to-day operational.

This boundary protects the strength of the role. When Enterprise Architects are embedded only within a single project, their perspective can narrow. When they take on pure management duties, architectural oversight can erode. Clear roles keep a long-term, cross-functional view in place.

That said, effective Enterprise Architects stay closely engaged with delivery. They attend key design sessions. They review solution proposals. They offer guidance when teams run into limits. The difference lies in purpose: they participate to maintain alignment, not to control every design decision.

A Definition Grounded in Output

What, exactly, do Enterprise Architects produce?

They build and update the core artifacts that map how the company is put together, show where it's headed, and set guardrails for technology choices. They run a steady, repeatable process that ties big-picture strategy to day-to-day delivery. They help leaders make smart calls across business and IT. They also keep multiple efforts aligned over time, across quarters, roadmaps, and long-term bets, so the whole system stays aligned with the vision.

This perspective opens the door to a more practical question: if a role is defined by what it produces, which outputs actually earn their keep? Plenty of diagrams look impressive but clarify nothing. Plenty of documents get written and ignored. So in the next chapter, we'll focus on the artifacts that truly matter, the ones people turn to when making decisions. These are the deliverables that shape conversations, create shared understanding.

CHAPTER 17

Artifacts That Are Really Used

Enterprise Architecture can produce alot of documents including diagrams, standards, roadmaps, models and governance write-ups. Shared drives and team sites swell with files until no one remembers what half of them are for. Judge the function by document volume alone and you'd swear the organization was world-class.

However, if you sit in on a steering committee or watch a budget debate unfold. Only a small stack of those files ever gets opened. Fewer still shape an actual decision. The rest sit untouched. The difference between what architects *could* create and what leaders *actually* rely on is often enormous, and usually ignored.

Learning to see which artifacts drive choices and which ones just decorate a repository is a practical skill. Maybe the most practical one an Enterprise Architect can build. Because what's the point of producing documentation if no one uses it?

This chapter deals with what actually happens in practice. Not the ideal archive of every possible artifact, but the handful organizations genuinely depend on, and how to keep those materials aligned, up to date, and reliable because if people don't trust them, they won't use them.

The Myth of the Complete Repository

Most architecture frameworks outline dozens of possible deliverables. There are models for every domain, matrices for every relationship, and diagrams for every layer of abstraction. An ambitious architecture team could easily spend years building a complete repository but few organizations can sustain that level of effort. Even fewer actually need to.

Most architecture teams rely on five to ten core artifacts. That's it. A capability map. One or two views of the current state. A picture of the future state. A roadmap. An application portfolio. A short list of standards. The rest remain half-finished, outdated, or never created at all. Some only appear for big programs then are forgotten.

Here's why. Artifacts live only when they shape real choices. If a document doesn't affect budget calls, priority debates, risk calls, or design sign-offs, it is effectively archived. Architects might admire the polish of the document but the business won't care.

The point isn't to crank out every artifact you can. It's to pick the few that matter and then use them again and again, where decisions actually happen.

What Organizations Actually Use

In most industries, a handful of artifacts keep showing up because they answer the same practical questions again and again.

Senior leaders typically ask: What can we do well right now? Where are we exposed? What are we funding next? To answer that, teams produce capability maps and roadmaps that make strengths, gaps, and planned investments visible.

On the technology side, the questions are different. Leaders want to see which applications are in place, where duplication exists, and which platforms are officially approved. Those needs give rise to application portfolios and standards catalogs.

Meanwhile, project teams are trying to place their work in context. They need to see how a proposed solution connects to the broader environment, which is why current-state and future-state diagrams, along with detailed solution views, become part of the mix. These artifacts are tools for making trade-offs. They shape budget choices, vendor selection, and project sequencing. When an artifact becomes part of a standing agenda, reviewed quarterly or cited in every major proposal, it earns its keep.

I once worked with an organization that had more than 200 architecture documents in a repository. Less than ten ever came up in governance meetings. When we looked at the rest, we found detailed data models for systems already retired, integration diagrams for projects finished years earlier, and standards no one followed. The repository looked impressive but he decision-making process didn't rely on it.

The lesson was simple: count what gets used, not what gets stored.

The Inconsistency Problem

You can pick the right artifacts and still lack clarity. The problem is unevenness. One capability map sticks to clean functional terms; another blends processes with technology platforms. One roadmap lays out three years in neat quarterly blocks; another stretches across five years with fuzzy checkpoints. Even something as basic as "lifecycle status" can mean different things depending on who put the list together.

That kind of misalignment erodes confidence. When leaders can't line up formats or definitions and see how they match, they stop using the material. Why bother reviewing diagrams if every one speaks a different language? Instead of debating the actual choice in front of them, architects waste time walking everyone through what the picture is supposed to mean.

Agree on how you name things, how far ahead plans should look, which colors signal what, and how deep you go into detail. A roadmap built this spring should resemble one built next year in its structure and design. A capability described in finance should follow the same pattern as one in operations: same structure, same level of detail.

Maintenance Is the Real Work

Building an artifact feels rewarding but keeping it current is a bit less fulfilling. However, whether that artifact stays useful over time depends far more on maintenance than on the initial build.

Take the application portfolio. The first version usually demands heavy lifting, gathering system names, identifying business owners, documenting technical stacks, mapping integrations, and noting lifecycle status. Once it's published, it anchors rationalization conversations. Then, before long, three new applications appear, two are retired, and ownership shifts. If updates don't follow, then confidence in the portfolio will erode because the contents can't be trusted.

Roadmaps behave the same way. A 24-month plan can look solid the day it's approved, but when budgets tighten or priorities shift, it becomes out of date. People may still point to it in meetings because it's the only thing on the slide, yet real decisions happen elsewhere.

A practical approach is to tie artifact updates to existing governance rhythms. For example, update the roadmap after each quarterly funding review. Refresh the application portfolio before annual planning. Revisit capability maturity assessments during strategy sessions. When maintenance is built into the cadence, it becomes predictable instead of optional.

Depth vs. Usability

Completeness and ease of use often pull in opposite directions.

System designers look for detail while executives look for something they can grasp quickly.

Picture a future-state architecture diagram. You could show the whole enterprise at a high level, zoom into domain components, map out integration flows in detail, or even add data movement paths. All of these views matter. But does every stakeholder need to see every layer? A diagram filled with too much detail might be perfectly correct, and still useless in a meeting.

The truth is that the artifacts people actually use are the ones they can understand without effort. They leave out the noise and speak in terms the audience already knows. They focus on the questions that matter most and skip the rest.

This isn't about dumbing things down; it's about targeting the audience. An executive roadmap shouldn't need a technical interpreter, and a technical standard shouldn't sound like it came from the marketing team. When a document misses its audience, it ends up untouched and forgotten.

Start by deciding what each artifact is for before you create it. Who's going to use it? In which meeting? To reach what decision? If those answers aren't clear, pause there, because a document without a defined role rarely gets used at all.

Avoiding the Archive Trap

Many architecture repositories eventually turn into archives instead of living tools. Files are stored, version numbers tick upward, diagrams labeled "final." But it's often hard to tell which artifacts are current, which are outdated, and which ones truly matter. Avoiding this trap takes clear ownership and lifecycle management. Each core artifact should have a named owner, often a domain architect, and a set review schedule. Older versions should be archived on purpose, not left sitting next to active documents.

Some organizations keep a simple dashboard that lists the core artifacts, their last update date, and their next planned review. This small step toward transparency builds accountability. If the capability map hasn't been reviewed in 14 months, everyone sees it.

You can buy the best platform on the market, loaded with features and dashboards, but it won't save you if no one maintains it. Clear rules, steady oversight, and people who actually follow them matter far more than the software. A simple shared drive, managed with care and consistent standards, will beat an expensive repository every time if the latter just sits there collecting dust. What good is a high-end system if nobody tends it?

Fewer Artifacts, Greater Impact

The truth is that a smaller, well-kept set of artifacts beats a sprawling but uneven library every time. That doesn't reduce the intellectual depth of Enterprise Architecture. It simply reflects how organizations actually work. Leaders are pressed for time, and they go back to tools they recognize, trust, and see tied directly to results.

In the Kettera approach, the focus is on kinetic value, artifacts that move the organization forward. A roadmap that drives investment sequencing. A capability map that exposes duplication. An application portfolio that informs decommissioning decisions. These are living tools, not just documentation.

Architects sometimes worry that limiting artifacts will weaken the discipline but the opposite is true. When a handful of artifacts are consistently used, the architecture function gains visibility and credibility. Conversations become structured and trade-offs become transparent. The architecture team is seen not as document producers but as partners in decision-making.

Building a Consistent System

A steady way of working doesn't happen by accident; you have to shape it. Figure out which artifacts your team truly relies on when making decisions. Don't send out a survey asking what people *like*: watch what they actually use. Sit in meetings. Scan the agendas. Notice which files get opened and which ones sit untouched. What shows up on the screen when decisions are made?

Standardize the basics so everyone plays by the same rules. Lock in shared templates and plain-English definitions. Set clear time frames for roadmaps so "long term" doesn't mean five different things. Spell out how applications move through their lifecycle. Choose simple, consistent names for capabilities and write all of this down in a short, practical guide people will read.

Make upkeep part of the rhythm. Connect updates to governance meetings so they aren't optional. Name a single owner for each artifact with no vague group responsibility. Put review dates somewhere visible. Add a quick check-in to your regular reporting cycle. Step back now and then to look for changes in the landscape. Tools that once felt essential can fade as the company grows, and new priorities may demand something different. When strategy shifts, your system should too. Let the process change when it needs to but never by accident or neglect.

A Practical Reflection

If you audited your architecture practice today, which artifacts would survive if you removed everything not referenced in the last six months? That thought experiment can be surprisingly revealing. It reminds me of a home decluttering exercise where you pickup items you are organizing and if they don't bring you "joy" or value, you get rid of them. Then you are left with what is really important.

Some teams discover that only the roadmap and application inventory remain. Others find that standards documents are frequently cited, while conceptual models are largely ignored. The goal isn't to judge, it's to understand. Once you see which artifacts actually carry weight, you can focus on strengthening them.

Enterprise Architecture succeeds when it influences behavior. Artifacts are simply a tool to assist with that influence. The ones people truly use shape conversations, funding decisions, and design choices. The rest are optional or noise.

In the next chapter, we'll look at one of the most widely used artifacts in Enterprise Architecture: the capability map. It often becomes the anchor for strategic alignment, investment analysis, and transformation planning. Used well, it drives action.

Understanding how to build and apply it effectively is essential to creating an architecture practice that produces artifacts not just for documentation, but for action.

CHAPTER 18

Capability Maps

Enterprise Architecture generates alot of documents, charts, and models. While many of them never earn their keep, a well-built capability map does. It grounds conversations between business and IT and gives everyone a shared frame of reference.

A capability map only works when people treat it as a living tool rather than a branding exercise. Clarity is everything as vague labels produce vague thinking. Clear definitions drive better decisions regarding priorities, funding, and change.

The Kettera Architecture Methodology treats capability maps as core tools. They actively link business goals, spending choices, application portfolios, and delivery roadmaps, tying strategy to execution in a way few other artifacts can. Before getting into how to build one, we should slow down and ask the obvious question: what do we really mean by a capability?

What Is a Capability Map?

A capability is something an organization needs to be able to do to reach its goals. It is not a process, a department, or a particular system. Instead, it describes, at a stable, outcome-focused level, the business ability itself.

Take "Customer Relationship Management." That is a capability. So are "Accounts Payable Processing" and "Order Fulfillment." Each one names an ability the organization requires, no matter how it is carried out, which team is responsible, or what software happens to support it.

A capability map lays out these abilities in a clear visual structure. It is usually arranged in levels. The top tier includes broad areas like Marketing, Operations, Finance, and Information Technology. Under each of these sit more specific capabilities, broken down into increasing detail. Together, they show everything the enterprise must be able to do.

The defining characteristic of a capability map is its relative stability. Processes change. Systems get replaced and reporting structures are reorganized. Through all of that change, capabilities endure. An organization that sells products must always be able to manage products, price them, sell them, deliver them, and collect payment. How those abilities are carried out may shift dramatically over time, but the capabilities themselves stay the same.

This stability makes capability maps especially powerful as architectural tools. They offer a consistent way to evaluate change.

What Are Capability Maps Used For?

Capability maps have real, everyday uses inside Enterprise Architecture. They help people make choices and ground strategy in reality. Leaders set big goals such as enter new markets, cut costs or boost customer loyalty, and those goals have to land somewhere. Which capabilities need to get better? Which ones are weak? Where should money go? The map forces those questions out into the open.

Second, capability maps strengthen portfolio management. You can link applications, technologies, and projects directly to the capabilities they support, which makes relationships easier to see. When five different systems back the same capability, overlap is hard to miss. On the other hand, if an essential capability has very little technology behind it, the opportunity becomes clear. The map highlights both duplication and gaps at a glance.

Third, capability maps provide a structured way to assess maturity. Teams can rate each capability using a scale such as ad hoc, developing, defined, or optimized, whatever model fits the organization best. Leaders quickly see where the company is solid and where it is vulnerable. Rather than arguing over which department should receive funding, discussions shift toward which capabilities need to improve.

Fourth, capability maps improve communication across groups. Business leaders often find technical architecture diagrams difficult to read, while developers may struggle to connect with high-level strategy slides. A clear capability map bridges that divide. It presents business priorities in plain terms without getting lost in operational detail.

Finally, teams use capability maps to rank initiatives. When a new project is proposed, architects trace it to the capability it strengthens. If that capability is strategically important and currently weak, the proposal carries more weight. If it targets an area that is already mature and low priority, it naturally drops lower on the list.

In short, capability maps connect strategy, architecture, and execution.

How Capability Maps Are Created

A capability map is not something you sketch on a whiteboard in an afternoon and declare finished. It takes care, focus, and a shared understanding of what truly counts as a capability, and what does not. Start with scope. Are you charting the whole organization or just one division? Trying to capture everything in one sweep usually buries the team in detail. Many companies begin with broad, top-level areas and fill in the specifics later, building the map piece by piece.

You also need to pinpoint the main capability domains. Drawing on the business architecture discussed earlier in this book, organizations often group their highest-level capabilities into areas such as:

- Strategy and Governance
- Marketing and Sales
- Product or Service Management
- Operations
- Finance

- Human Resources
- Information Technology

These broad categories give you a useful framework, but they shouldn't be considered prescriptive. The map needs to mirror how the organization truly works day to day and not a theoretical vision.

Once domains are defined, capabilities are broken down into levels. For example, under Marketing and Sales, you might identify:

- Market Analysis
- Campaign Management
- Lead Management
- Opportunity Management
- Pricing Management
- Customer Relationship Management

Each capability can be divided into smaller pieces. Take "Campaign Management." It could cover campaign planning and performance measurement. How far you break it down depends on why you are building the map in the first place. Too high-level, and it's vague. Too detailed, and it becomes noisy.

Use nouns that describe what the organization is able to do, not action phrases. "Invoice Processing" is clear and steady. "Process Invoices Quickly" is not. The first names an enduring ability. The second adds a speed target that is left to interpretation and could change next year.

Workshops with business stakeholders help surface and confirm capabilities. These sessions need strong facilitation. People will start talking about workflows or specific tools and that's natural. The architect has to keep pulling the group back to the core question: What must this organization be able to do, no matter how it gets done?

After drafting the map, check it against both strategy and day-to-day operations. Look for big gaps. Watch for overlaps. Adjustments are part of the process. Capability maps should grow and improve over time, but rebuilding them every quarter defeats the purpose.

Examples of Capability Maps in Practice

Take a mid-sized retail company as an example. At the top level, its capability map could include areas such as:

- Product Management
- Supplier Management
- Merchandising
- Store Operations
- E-commerce Operations
- Customer Management
- Financial Management
- IT Services

Each of these represents a core business ability rather than a department name.

Within Store Operations, the map would typically break down into more specific, second-level capabilities, including:

- Inventory Management
- Point of Sale Processing
- Staff Scheduling
- Store Compliance
- Loss Prevention

Each of these can be evaluated for maturity. For example, Inventory Management might be highly optimized, backed by integrated systems and automated replenishment. Staff Scheduling, by contrast, may still depend on spreadsheets and manual coordination. This clear view guides investment discussions.

Now imagine the organization plans to grow its e-commerce presence. The capability map highlights E-commerce Operations and the capabilities that support it, such as Online Catalog Management, Digital Payment Processing, and Fulfillment Integration. Architects can quickly see which capabilities need to be strengthened to support the expansion. Technology investments are no longer arbitrary. They are tied to specific business abilities.

A financial services firm offers another example. Top-level capabilities might include:

- Client Onboarding
- Risk Assessment
- Portfolio Management

- Regulatory Reporting
- Claims Processing

New reporting rules from regulators hit Regulatory Reporting first and may also touch Risk Assessment. That's where the pressure is felt. A capability map makes this obvious. Architects don't have to comb through every system across the company. They pinpoint the capability involved, then follow the thread to the applications and data that support it.

These maps should be visual. Capability maps are often color-coded so the story jumps off the page. Heat mapping is a popular method. Each capability gets shaded to reflect maturity, risk, cost, or strategic priority. A red-to-green spectrum can spotlight weak performance while deep blue might signal heavy investment; money pooling in one place. With that added layer, the map becomes a working decision tool rather than a flat picture on a slide.

Common Pitfalls

Capability maps are helpful, but they're easy to get wrong.

A common error is mixing up capabilities and processes. Processes show how work moves from one step to the next. Capabilities describe the ability to carry out that work. When the two get blurred, the map swells into something that looks more like a process flowchart than a capability view.

Teams also tend to anchor capabilities to the current org chart. If you see "Marketing Department" listed as a capability, that's a red flag. Departments get renamed, split, and reshuffled. Capabilities should stay steady through those changes.

Overengineering is just as much of a problem. Some teams try to break capabilities down to extreme detail, even to individual tasks. The result is a map that is technically accurate but hard to use in practice. The right level of detail helps people make decisions without burying stakeholders in too much information.

Capability maps must actually be used. If they are created once and never used in project reviews, funding talks, or strategy planning, they lose credibility. The artifact earns authority only through consistent use.

Capability Maps in the Kettera Methodology

In the Kettera Architecture Methodology, a capability map connects Business Architecture to Technical Architecture. It turns big-picture goals into clear, manageable capability areas that systems can actually support through applications, data, integrations, platforms, and even core processes. Instead of vague strategy talk, you get concrete areas the technology stack can act on.

Architects start with capabilities. They ask which ones need to get better over the next 12 to 24 months and how much progress the business expects to see. Is the goal incremental improvement or a serious leap forward? From those answers, they shape initiatives, decide what comes first, and tie each effort to the right supporting technologies. The roadmap grows out of capability priorities, not the other way around.

Capabilities act as anchors. They don't swing wildly year to year, even when strategy shifts or new tech trends grab attention. Because they stay fairly steady, they create consistency across planning cycles, acting as a stable reference point when priorities move around. Annual plans may change. Technology fads come and go. The capability map keeps architecture discussions rooted in what the business actually needs and not what's trendy this quarter.

Key Takeaways

A capability map shows what an organization needs to be able to do. It describes abilities, not processes, org charts, or IT systems. What makes it useful is that it stays relatively stable over time and makes those abilities easy to see at a glance.

Teams rely on capability maps to link strategy with day-to-day work, manage project portfolios, gauge maturity, and decide where to invest. Built with clear names, consistent levels of detail, and input from stakeholders, they serve as working tools, not just polished diagrams that sit in a slide deck.

The strength of a capability map becomes clear when change happens. Expansion, regulation, cost pressure, innovation; each can be viewed through the lens of the capabilities they affect. That perspective helps guide architecture decisions.

In the next chapter, we build on this foundation by exploring Current vs. Future State Diagrams. If the capability map defines what the enterprise must be able to do, current and future state views show how well those abilities are supported today, and how they need to evolve tomorrow.

CHAPTER 19

Current vs Future State Diagrams

Architecture exists to drive change. Companies do not pour money into enterprise architecture just to create tidy records of what they already have. They invest because something isn't good enough. Costs are too high. Delivery takes too long or customers complain. At its core, improvement is the goal.

Current-state and future-state diagrams make that change concrete. They show how things work today and how they should work tomorrow. Without that contrast, it can be difficult to see what you are actually improving? A diagram of the present captures reality. A diagram of the future lays out a vision, shaped by strategy and business priorities. Side by side, they expose the gap.

These visuals are practical tools leaders use to understand their position, define their direction, and map the ground between the two. In the Kettera EA Methodology, they connect big-picture strategy to real delivery plans. You cannot design a sensible transition without seeing both endpoints clearly.

What Are Current-State and Future-State Diagrams?

A current-state diagram shows how a business capability, process, application environment, data flow, or technical platform actually runs today. It answers a straightforward question: what really happens right now? Not the way it was meant to work. Not the polished version but the messy, real-world version.

These diagrams usually map systems, integrations, manual steps, data stores, decision points, and sometimes who owns what. Picture a customer onboarding flow: you might see a CRM, an identity verification vendor, a billing platform, and a manually updated spreadsheet holding everything together. You might also find batch file transfers, approval emails bouncing between inboxes, and the same customer data entered twice.

A future-state diagram shows where you want to go instead. It lays out how that same capability or domain should operate once planned changes are in place. This is not a blue-sky sketch. It's a designed target that fits the business strategy, real-world limits, and agreed architectural standards.

Where the current state may show fragmented systems and manual workarounds, the future state can illustrate consolidated applications, real-time integrations, standardized data models, and automated approval workflows. It offers a clear visual answer to a simple question: what are we building toward?

Both diagrams can exist at different levels of detail. At a high level, they may show major business capabilities and core systems. At a more detailed level, they can outline specific APIs, data flows, or infrastructure components. The right level of detail depends on the audience and the purpose.

What Are They Used For?

Current-state and future-state diagrams have real, practical value inside enterprise architecture. They build shared understanding. People across the business picture the same system in very different ways. Operations may call a process smooth, while IT sees brittle integrations and years of tech debt. Finance tracks cost centers. Product teams obsess over customer touchpoints. How can teams move together if they're imagining different realities? An accurate current-state diagram pulls those views into one clear, honest snapshot.

When you map systems, processes, and data flows side by side, the weak spots tend to jump out. Redundant applications clutter the landscape. Manual rework shows up as looping steps. Integrations that depend on one person look dangerous. Shadow spreadsheets lurk in the corners. Put it all on paper and the problems get hard to defend.

Future-state diagrams do something different. They point everyone at the same destination. Without that shared target, work gets off track. One team upgrades an app while another builds a workaround for it; wasting time and money. A future-state view cuts through that noise. It becomes the anchor for decisions and trade-offs, a simple reference that says, "This is where we're going."

These diagrams also guide investment decisions. When executives review funding requests, they can test each proposal against the future-state model: Does this initiative move us toward the target architecture, or does it keep us anchored in the current one? With the architecture clearly laid out, prioritization shifts from opinion to evidence.

Current- and future-state diagrams also ground transformation roadmaps. They mark the two endpoints the roadmap must connect. The differences between them, across functions, technology, and organization, form the real work of the transition plan.

How Are Current-State Diagrams Created?

Creating a current-state diagram starts with discovery. Architects collect information from system documentation, interviews, workshops, and sometimes direct observation. They ask direct questions: Where does this data come from? Who interacts with it? What happens when the system fails? How often is this integration fixed manually?

Rarely does a single source provide the full picture. Documentation is often outdated. People tend to describe how things are "supposed" to work. The architect's job is to pull these inputs together into a clear and accurate picture. This usually means reviewing drafts with subject matter experts and refining the diagram until everyone agrees.

Scope discipline is critical. Trying to capture the entire enterprise in one diagram creates clutter and confusion. Keep it focused. Architects define a boundary, such as customer onboarding, order fulfillment, or financial reporting, and stay within it. If needed, they create separate diagrams for related areas.

Notation matters, but not too much. The goal is clarity, not artistic polish. Boxes for applications, arrows for data flow, and swim lanes for organizational responsibility are typically enough. Consistency across diagrams is more important than fancy symbols. In the Kettera approach, standard visual conventions are set early to keep each architect from inventing their own style.

Here's a concrete case. Imagine a mid-sized retailer mapping out how it actually handles orders from start to finish. The "current state" diagram could include:

- An online store that collects customer orders
- Staff manually exporting order details into the warehouse system
- A separate inventory tool that updates once a night through a batch run
- Customer emails sent one by one using saved templates
- Financial records typed into the ERP system at the close of each day

Put all of that on one page and the cracks jump out. The same data gets typed more than once. Inventory numbers lag behind reality for hours. No one planned it that way; people just worked around the friction because that's how it had always been done.

But once you see the full chain, the challenges become hard to ignore.

How Are Future-State Diagrams Designed?

Designing a future-state diagram is more than tidying up what already exists. It means aligning the design with strategy, constraints, and the standards defined earlier; organizational goals, technology guidelines, and architectural principles all come into play. It usually starts with a clear definition of business objectives. Imagine a retailer aiming for real-time inventory visibility and same-day shipping on selected products. Those goals quickly shape the architecture. Batch jobs might give way to real-time APIs, separate inventory systems could be merged, and automated alerts may take the place of manual notifications.

Standards and reference architectures are applied next. When a company has already chosen a particular integration platform or cloud provider, the diagram needs to reflect that decision. Enterprise architecture acts as a set of guardrails here. The vision can be ambitious, but it still has to fit within budget limits and the organization's capacity to execute.

Trade-offs inevitably surface. While a full system consolidation might look ideal on paper, funding realities often call for phased modernization instead. For that reason, future-state diagrams typically aim at a practical target within a defined window, often tied to a 12- to 24-month roadmap rather than an aspirational five-year horizon.

Returning to the retailer example, a future-state diagram

might show:

- E-commerce platform integrated with warehouse management through real-time APIs
- Unified inventory database available to both systems
- Automated customer notifications triggered by shipping events
- Direct integration between the order system and ERP, so staff no longer re-enter data

The future-state model simplifies flows and cuts manual work. It ties the technical design to the strategic goal of faster fulfillment.

Comparing Current and Future States

Side by side, the diagrams tell the real story. What looked vague before suddenly becomes obvious.

In the present view, people might handle tasks by hand; in the future view, those steps are gone. Two tools that once did the same job might be merged into one. Information that used to move in overnight batches could start flowing instantly. Even ownership can shift from one team to another. Every one of these gaps means effort such as configuration, coding, migration, training, or shutting old systems down.

Architects often mark up the diagrams to show what each change means. An old platform in the current picture might carry a bold “Retire” label. A brand-new connection in the future version could be tagged “Build.” These simple notes turn drawings into action. They become the skeleton of the transformation plan.

When both the current and future states are visible, stakeholder discussions take on a different tone. The conversation shifts away from arguing about whether modernization is needed and toward more practical questions: How do we get from here to there? In what order should we act to limit disruption? Which dependencies need attention first and which can wait?

Common Pitfalls

Despite their usefulness, current- and future-state diagrams can lose their impact if handled poorly.

One common mistake is creating a current-state diagram that is too polished. If architects leave out manual workarounds or shadow systems because they seem informal, the picture will mislead decision-makers. Accuracy matters more than elegance.

Another error is designing a future state that is disconnected from reality. If the diagram assumes unlimited funding, perfect data quality, and instant organizational change, it will not stand up to real-world scrutiny. A credible future state pushes the organization forward but does not ignore its limits.

Too much detail kills clarity. When a diagram is filled with tiny elements and side notes, people stop looking at it. The whole point is to get an idea across, not to show everything you know. If someone can't grasp what the drawing says in a few minutes, something's off. Strip it back or break it into smaller pieces, whatever makes it easier to follow as clarity always beats cleverness. Overly complicated visuals don't impress stakeholders, they push them away.

From Diagrams to Roadmaps

Current- and future-state diagrams mark two fixed points. Everything between them is the transformation journey, planned, funded, and governed step by step.

From there, the next tool in the enterprise architect's kit comes into play: the roadmap. If these diagrams answer "Where are we?" and "Where are we headed?" the roadmap tackles "How do we move and when?" The following chapter breaks down how to build 12- to 24-month roadmaps that turn architectural intent into coordinated execution.

CHAPTER 20

Roadmaps (12 to 24 month)

Current-state models show where the company is today. Future-state diagrams sketch out where it plans to go. They help communicate the intent but they don't answer the questions: *How do we actually get there and how long will it take?* A roadmap answers that.

In the Kettera EA Methodology, the 12–24 month roadmap stands out as one of the most visible and frequently used architecture artifacts. It translates architectural vision into a set of sequenced initiatives with clear time frames. Unlike long-range strategic plans that can become vague, this roadmap works within a practical window, far enough ahead to shape meaningful change, yet close enough to influence real funding and delivery choices.

What Is a Roadmap?

A roadmap is a time-based view of change initiatives that move the organization from its current state to its defined future state. It links capabilities, applications, data, and technology investments into a coordinated plan across a set time horizon, typically 12 to 24 months. A roadmap is not a project plan. It does not include detailed task breakdowns, sprint plans, or resource assignments. Instead, it works at a higher level. It answers questions such as:

- Which major initiatives will take place this year?
- In what order should they happen?
- What dependencies exist between them?
- How do they align with strategic goals?

A well-built roadmap tells a story. It shows how modernization efforts build on one another, how risks decrease over time, and how business value grows in measurable stages. When done well, it lets executives see progress across quarters and across products, not just in isolated projects.

A 12–24 month window isn't random; it's chosen on purpose. A single year isn't always enough runway to pull off real architectural change, as larger transformation takes time. If you look past two years, though, you will be guessing due to the volume of changes that will take place in the business, political and social environments.

Two years hits the sweet spot between bold and practical. In that span, you can redesign foundations, test new systems, and actually ship improvements without betting on a unknown future. Go longer and things will move under your feet, technology advances, budgets get cut or reshuffled and leadership changes its mind. What looks solid today can feel outdated in 18 months. The

12–24 month range keeps ambition alive while staying grounded in what's likely to hold true.

What Are Roadmaps Used For?

Roadmaps are created with several audiences in mind. Senior leaders look to them to understand how investments are sequenced and what trade-offs are involved, while portfolio managers use them to sync plans with funding cycles. Delivery teams check them to see upstream and downstream dependencies, and enterprise architects rely on them to keep initiatives technically aligned across the organization.

At the strategic level, a roadmap makes direction visible. Executives can quickly see how initiatives connect to business capabilities and goals. Rather than sorting through isolated proposals, they view a coordinated progression, starting with foundational platform upgrades, moving to integration consolidation, and eventually introducing customer-facing innovation.

Sequencing becomes clear as well. An organization can't roll out advanced analytics before improving data quality and integration. By laying out initiatives in a deliberate order, architects discourage teams from chasing appealing but premature ideas. The roadmap shows why some projects need to happen before others.

Budget planning becomes more practical when it's backed by a clear roadmap. Funding discussions shift from "Should we approve this project?" to "Is this initiative in the right place in the sequence?" The roadmap frames the conversation around architectural progress rather than stand-alone return-on-investment calculations.

Finally, roadmaps create accountability. When initiatives slip, the impact shows up across the timeline along with the impact to dependencies. If a core data platform is delayed, downstream analytics programs have to adjust. That visibility drives proactive management instead of reactive crisis response.

How Roadmaps Are Created

A roadmap is the next logical step after the work that has already been done in previous stages. It grows from capability maps, side-by-side views of today versus tomorrow, and the guidelines set by your architecture.

Architects spell out what the future should look like across the business, applications, data, integration, and the underlying platforms. They make the target concrete. Then they compare that picture to what exists now and study the distance between the two. What's missing? Which systems have outlived their usefulness? Where do new standards need to be put in place and enforced?

Every gap turns into action. Sometimes one gap becomes several initiatives. Those proposed efforts are then sized up and compared, usually based on impact, effort and risk. Which ones move the needle? Which ones will cost time and money without much payoff?

- Strategic alignment: Does this initiative clearly advance stated business goals?

- Dependency sequencing: Must another effort be completed before this one can begin?
- Risk and technical debt reduction: Will it eliminate instability or reduce fragmentation?
- Investment size and the availability of people and funding.

Sequencing is seldom a straight line. Dependencies often stretch across multiple domains. Consolidating applications, for instance, may only be possible after data has been standardized, and a cloud migration can hinge on prior network upgrades. Architects need to map these connections with care; if they don't, the roadmap risks becoming wishful thinking instead of something teams can actually deliver.

After sequencing, initiatives are distributed across a quarterly timeline. Roadmaps that span 12–24 months are typically organized by quarter rather than by month, since quarterly planning matches financial cycles and still leaves room to adjust when reality intervenes.

Then comes prioritization. Two years is never enough time to do everything. Trade-offs become unavoidable: some efforts move beyond the planning horizon, while others are pulled forward because they open the door to later gains. Architects guide these discussions, anchoring decisions in architectural consistency instead of individual preference.

The final step is validating the plan with stakeholders. Roadmaps are collaborative tools. Business leaders, IT leadership, finance, and delivery teams need to agree that the sequence is realistic and matches funding assumptions.

Example: Application Modernization Roadmap

Picture a mid-sized company juggling too many disconnected apps while maintenance bills keep climbing. Systems don't talk well to each other. Costs creep up every quarter. The target state replaces that sprawl with a cleaner ecosystem: fewer applications, shared integration standards, and a clear move to a cloud-first platform approach; built to scale and easier to manage.

A two-year roadmap lays out how to move from today's tangled environment to that simpler future. Over 24 months, the organization phases out redundant tools, aligns systems around common integration methods, and shifts core workloads onto a cloud platform. The goal is straightforward: reduce complexity, lower support costs, and create a setup that can actually grow with the business.

Quarter 1–2

- Complete a thorough assessment to rationalize the application portfolio
- Define integration standards and set up the API gateway
- Start implementing the cloud landing zone

Quarter 3–4

- Decommission five overlapping legacy systems
- Move the customer portal onto the cloud platform
- Roll out enterprise-wide identity and access management

Year 2, Quarter 1–2

- Bring together CRM systems into a single platform
- Make data models consistent across sales and marketing
- Extend API connections to key partners

Year 2, Quarter 3–4

- Shut down on-premise data center infrastructure
- Roll out advanced analytics tools
- Improve cloud cost management processes

Notice the sequence. The organization does not roll out advanced analytics in Year 1. It first standardizes integration, consolidates systems, and modernizes infrastructure. Each initiative builds on the work before it. The roadmap shows clear progression, not a set of isolated activities.

Example: Business Capability–Driven Roadmap

Roadmaps don't have to revolve around systems and tools. They can be built around what the business is actually trying to get better at. Picture a company aiming to deepen customer relationships, run leaner operations, and make its numbers easier to understand and trust. Here, the plan is organized by capability, not by technology; a subtle shift that changes the whole conversation. Why talk about software first when the real goal is stronger performance?

Customer Engagement

- Launch a unified, multichannel communication platform (Year 1 Q1–Q2)

- Connect the CRM to marketing automation tools for smoother campaigns (Year 1 Q3–Q4)

Operational Efficiency

- Automate the order processing workflow to cut manual effort (Year 1 Q2–Q3)
- Roll out predictive demand forecasting to reduce waste and stock issues (Year 2 Q2–Q3)

Financial Transparency

- Unify financial reporting systems (Year 1 Q1–Q3)
- Roll out an enterprise budgeting platform (Year 2 Q1)
- Launch real-time performance dashboards (Year 2 Q3–Q4)

Organizing the roadmap around core capabilities helps leaders see exactly how each investment ties to company goals. The technology work is still there, but it's described in terms of business results, not technology.

Common Pitfalls in Roadmapping

Roadmaps often fail, not because of bad intent but because the scope is unrealistic. One common mistake is filling the timeline with too many initiatives running at the same time. When every quarter includes a dozen major transformations, the roadmap stops being credible. Capacity limits need to be acknowledged openly.

Another issue appears when roadmaps turn into static documents. A 24-month plan will not stay relevant in a fast-moving organization. Review the roadmap quarterly and adjust sequencing as priorities shift or as initiatives finish earlier or later than planned. Direction can stay stable without locking the schedule in place.

Too much detail is just as problematic. A roadmap packed with minor tasks quickly loses clarity. It should spotlight major initiatives; those that meaningfully change architecture or capabilities. Leave detailed task tracking to program and project plans.

Architects often get enamored with technical purity. They polish the sequence until it shines; every dependency and layer perfectly arranged, yet they forget the messy parts of the business like budget timing, shifting priorities, whether the team is even ready to take it on. What good is a flawless design if no one can fund or run it? A roadmap only holds up when it balances clean architecture with the day-to-day facts of how the organization actually operates.

A plan that ignores money cycles or staff capacity will grind to a halt, no matter how good it looks on paper. The strongest roadmaps respect both the logic of the system and the practical limits of the people expected to deliver it.

Visual Presentation and Format

The strongest roadmaps usually take the form of a straightforward timeline divided into quarters, with each initiative shown as a bar stretching across its expected timeframe. Colors can signal different domains such as business, application, data, integration, or platform, so readers can spot patterns at a glance.

You can add light arrows to show dependencies, but resist the urge to turn the page into a wiring diagram. If it looks like a subway map, you've probably gone too far.

Design matters less than clarity. An executive should be able to scan the page and grasp what happens when, and why, within a few minutes. Keep the main view focused on sequence and intent, and place the detailed explanations in supporting materials where interested readers can dig deeper.

Alongside the visual timeline, include a brief narrative that explains the main themes for each year. For example: "Year 1 focuses on foundation and consolidation; Year 2 emphasizes expansion and optimization." This framing helps stakeholders see how individual initiatives fit into the bigger picture. It gives the timeline context.

Roadmaps in the Kettera Methodology

In the Kettera EA Methodology, a roadmap grows out of clear limits, the Three C's, locks into business architecture domains, and gets checked against technical standards. Strategy on one side and delivery on the other. The roadmap ties them together.

It feeds directly into the enterprise architecture rhythm described later in this book. Leaders use it to set priorities and map resources. When fresh ideas show up in the innovation pipeline, they're measured against what's already planned. Do they move things forward, clash with the direction, or force a reshuffle?

The roadmap earns its keep here. It becomes the go-to reference for decisions across the company, a shared anchor that shapes choices long before projects begin.

Key Takeaways

A 12–24 month roadmap turns architectural vision into a clear sequence of actions. It links future-state goals to concrete initiatives within a realistic planning window, showing how ideas move from concept to execution. Rather than acting as a simple project list, it presents a connected narrative of change.

Strong roadmaps start with a clear understanding of current gaps and are organized around dependencies and timing. They are checked against available capacity to ensure the organization can actually deliver what is planned. Plans should stay flexible, with regular reviews and adjustments, and they focus on meaningful initiatives instead of breaking work into overly detailed task lists.

Above all, a roadmap makes architecture practical. Future-state diagrams on their own stay theoretical; a roadmap ties them to coordinated, measurable action across the enterprise.

In the next chapter, we will look at another key artifact: the Application Portfolio. While the roadmap shows how applications will change over time, the portfolio gives a clear inventory and grouping to understand what exists today, and why rationalization is often the first step in any transformation effort.

CHAPTER 21

Application Portfolio

Ask around in almost any mid-size or large company, "How many applications do we actually run?" You'll get wildly different answers. One leader will name the big, customer-facing platforms. Another will toss in team-built tools, forgotten legacy systems, random SaaS sign-ups, and the shadow IT nobody admits to owning. Almost no one can give a clean, confident number. That fog is the whole reason the application portfolio matters.

The application portfolio is one of the most hands-on tools in Enterprise Architecture. In the Kettera EA Methodology, the portfolio ties together business capabilities, technical design, and future roadmaps into something you can actually track and manage. It turns abstract thinking into a living inventory.

What Is an Application Portfolio?

An application portfolio is an organized record of every

software application an organization uses, along with key details that describe and evaluate each one. It's more than a simple inventory. The portfolio captures information leaders need to judge value, risk, overlap, and how well each system supports business goals.

In basic terms, it answers three questions: What applications do we have? What do they do? Who owns them? But a well-developed portfolio goes much further. It shows which systems are mission-critical, which business capabilities they enable, what technologies they depend on, where they sit in their lifecycle, what they cost, and whether they should be merged, modernized, or retired.

Picture a financial portfolio, only instead of stocks and bonds, you're tracking CRM systems, HR platforms, integration tools, homegrown apps, reporting solutions, and everything in between. Some represent long-term bets tied to strategy. Others are dependable workhorses that quietly keep operations running. And a few are aging systems that add risk if no one is paying attention.

What Is the Application Portfolio Used For?

Enterprise Architecture is about steering change, and you can't steer anything if you don't know what's already in place. The portfolio lays out the current landscape; every system, every owner, every dependency, so leaders can choose what to fix, fund, or cut with clarity.

Redundancy becomes more apparent with an Application Portfolio. Different teams often buy similar SaaS tools to handle similar work. Marketing grabs one survey app while HR signs a contract for another. Each choice looks reasonable on its own. Together, they add extra costs, messy integrations, scattered data, and confused reporting. A complete portfolio makes the overlap visible.

Aging is slow and less obvious than overlap. Applications don't become unsupportable overnight. Support can end without notice or the developer who built the custom logic could leave. The portfolio helps architects tag each system: strategic, tolerable, invest, migrate, or retire. Those labels, along with periodic review, help shape the roadmap and avoid surprises related to aging systems.

Risk management is another area that deserves our attention. A payroll or regulatory reporting system carries far more risk than a small team productivity app. When you document technical dependencies, hosting models, and support setups, the portfolio turns into a clear risk heatmap. You can quickly see which systems run on aging platforms, which rely on a single person, and which have no disaster recovery plan in place. Instead of guessing, you have visible answers.

Budget planning improves as well, when we implement a solid application portfolio. When Finance questions a rising technology budget , the portfolio makes the numbers easier to explain. Spending connects directly to business capabilities. Overlapping tools are visible and defensible. You can show how retiring older systems helps balance the cost of new investments, grounding architectural decisions in financial accountability rather than theory.

Finally, the portfolio supports strategy alignment. If the organization is shifting toward digital self-service, the portfolio should clearly show which applications support customer-facing digital channels and which are still manual or paper-based. Gaps and misalignments will stand out better.

How Application Portfolios Are Created

Building an application portfolio doesn't really create a fancy set of diagrams. It's mostly rolling up your sleeves and digging through messy information across multiple departments and teams. You collect data, talk to people, compare notes, fix inconsistencies, and keep asking questions. It's not glamorous work but it's important to capture as much of the ecosystem as possible.

Start by drawing clear boundaries. Are we just looking at company-wide platforms? Just tools that individual departments bought on their own? Every SaaS subscription tied to a corporate card? Go wide at the beginning to gain maximum visibility within the defined scope. Those "tiny" apps and integrations everyone shrugs off have a habit of causing trouble later, especially when no one remembers they exist.

The next step is data collection. You'll pull information from many places: procurement records, IT service management tools, cloud subscription reports, financial systems, and interviews with business leaders. It doesn't take long to spot problems such as different names for the same application, duplicate listings, even systems no one recalls approving.

After you identify an application, record it using a standard set of attributes so every entry is described the same way. Typical data fields include:

- Application name
- Business owner
- Technical owner
- Supported business capabilities
- User base size
- Hosting model (on-premises, SaaS, cloud IaaS/PaaS)
- Vendor and contract status
- Annual cost (licensing, support, infrastructure)
- Lifecycle status (strategic, maintain, retire, etc.)
- Integration dependencies
- Data sensitivity classification

Notice this isn't purely technical. The portfolio connects business and IT. An application without a clear business owner is a red flag while one without cost visibility points to governance gaps.

After the initial data collection, validation sessions are critical. Architects review the draft portfolio with business and IT stakeholders. They make corrections and confirm ownership. Applications once believed to be widely used may actually support only a small team. Others may prove far more critical than expected.

The final step is to layer in governance. An application portfolio is not a one-time effort. It must be kept current. New projects must register applications. Retired systems must be removed. Major status changes must be updated. In the Kettera methodology, this upkeep is built into the architecture cadence and project approval process. If a project introduces a new application, it isn't complete until the portfolio is updated.

Structuring the Portfolio for Insight

Most teams begin with a simple spreadsheet. As things grow, many switch to dedicated portfolio management software. The tool itself isn't the real issue. What actually counts is having a clear setup and the discipline to keep it accurate over time.

A practical way to organize the portfolio is by business capability. Don't sort applications alphabetically. Instead, place them under the work they support: Customer Management, Order Fulfillment, Financial Reporting, Human Resources, or Product Development. When you see systems grouped by what the business actually does, patterns jump out.

If three different tools sit under "Customer Relationship Management," that tells a very different story than a single, shared platform. Why are there so many? Are they solving different problems, or has overlap crept in unnoticed? Grouping by capability makes duplication obvious and sparks smarter questions about consolidation, cost, and complexity.

Another helpful way to think about the portfolio is through lifecycle categories. Many organizations use a simple four- or five-stage model to sort applications:

- Strategic (actively funded and aligned with the future direction)
- Invest (important, but in need of improvement or modernization)
- Maintain (stable, with no significant new spending planned)

- Tolerate (acceptable for now, though a replacement has been identified)
- Retire (scheduled to be decommissioned)

These categories shape real budget choices. If a system is tagged "Retire," it shouldn't be receiving major upgrade dollars. When something is marked "Strategic," the organization should protect it and update it when needed; not let it quietly decay.

Visualizations make things clearer. Heatmaps that map business value against technical health are especially useful. High value and low health? Make modernization a priority. Low value and low health? Retire it. High health but low value? Ask if it should exist at all.

Sample Application Portfolio Entry

Here's a stripped-down example.

Application Name: XYZ CRM

Business Owner: VP of Sales

Technical Owner: Director of Enterprise Applications

Primary Capabilities Supported: Lead Management, Opportunity Tracking, Sales Forecasting, Pipeline Reporting

User Base: 220 employees across the company

Hosting Model: SaaS (cloud-based, fully vendor-managed – no on-prem servers to babysit)

Vendor: XYZ Systems Inc. (under contract through 2028)

Annual Cost: $480,000

Lifecycle Status: Strategic

Integrations: ERP for order processing, Marketing Automation Platform, Customer Support System, Data Warehouse

Data Classification: Confidential — Customer Data

What does this snapshot really tell you?

It spells out who owns the system, who keeps it running, what the business relies on it for, how many people use it, what it costs each year, and how tightly it's connected to other core tools. You can see the risk exposure, the financial commitment, and the operational footprint at a glance.

Now look at a second entry and compare.

Application Name: Sales Tracker

Business Owner: Regional Sales Director (East)

Technical Owner: No formal owner assigned

Primary Capabilities Supported: Opportunity tracking (regional only)

User Base: 18 users

Hosting Model: SaaS (self-provisioned)

Vendor: Small independent provider

Annual Cost: $22,000

Lifecycle Status: Tolerate

Integrations: None

Data Classification: Confidential – Customer Data

The questions become a little more obvious. Why are two

separate systems tracking the same opportunities? How is there no formal technical owner? And why is sensitive customer data sitting outside enterprise integration and governance controls? Even without deeper analysis, the gaps are hard to ignore.

Common Pitfalls

Many organizations treat the portfolio as a compliance exercise. They build it to satisfy auditors or to claim architectural maturity. The result is a static document that no one actually uses. That defeats the purpose.

Another common mistake is making it too complicated. Collecting fifty attributes for each application may sound thorough, but if the data cannot be kept accurate, the portfolio quickly becomes outdated. Start with what drives decisions: ownership, cost, capability alignment, lifecycle, and risk indicators.

There is also a tendency to overlook shadow IT. Business units may subscribe to tools with corporate credit cards. Architects who ignore this or miss it due to lack of visibility create an incomplete picture. When the portfolio is positioned as a support tool rather than a policing mechanism, participation among these business units improves and they are more likely to discuss their toolsets and remove these gaps.

Connecting to Other Artifacts

An application portfolio never exists by itself. It ties straight into other core Enterprise Architecture artifacts and makes them usable instead of abstract.

Capability maps spell out what the business must be able to do. The portfolio makes it clear which systems actually deliver that work. Put the two next to each other and blind spots jump out, along with redundant tools doing the same job.

Current- and future-state diagrams map how systems connect and exchange data. The portfolio supplies the hard inventory behind those pictures such as the names, owners, and roles that make the drawings real. Roadmaps lay out what will change and when. Lifecycle tags in the portfolio shape those plans and show what gets upgraded, replaced, or retired.

Key Takeaways

An application portfolio is a structured, maintained inventory of every application used across the enterprise or a scope within that enterprise, paired with business, technical, financial, and lifecycle details. With this information in one place, leaders can make smarter choices about where to invest, what to consolidate, which systems to modernize, and what to retire. It also reveals duplicate tools, makes ownership visible, and brings hidden risks into view.

Building the portfolio takes steady data collection, careful validation, and clear governance to keep it current. When tied to the Kettera EA Methodology, it strengthens capability alignment, informs roadmap planning, and improves financial transparency.

In the next chapter, we move from individual applications to the wider landscape of standards and reference frameworks. If the application portfolio shows us what we have, standards and reference frameworks explain how those applications should be designed, integrated, and governed from here on.

CHAPTER 22

Standards and Reference Frameworks

Enterprise Architecture can create a lot of deliverables, and plenty of them have nothing to do with implementing specific changes. Some are there to keep things from spinning out of control. Standards and reference frameworks sit in that camp. They don't map to one program or highlight a single effort. They define the guardrails, shared expectations, and preferred ways of doing things that everyone is supposed to follow.

Enterprise Architects put standards and frameworks in place to cut friction, boost consistency, speed teams up, and protect alignment across the company. The goal isn't control for its own sake. It's smoother delivery, fewer surprises, and work that fits together instead of fighting itself, which prevents headaches later.

What Are Standards?

Standards are written rules that spell out what's acceptable inside an organization. They lay out which technologies are approved, how systems should be designed and connected, how data should be named, and what security controls are required. When someone asks, "Which databases can we use?" or "How do we secure an API?" or "What naming pattern should this field follow?" the answers should already exist in a standard.

Some of these rules are mandatory. For instance: "All external-facing applications must use multi-factor authentication." Others serve more as guidance, such as recommending certain programming languages for new projects. How strictly they're enforced varies by governance model, but in every case, the aim is to limit avoidable differences and the confusion that comes with them.

Enterprise standards are not meant to stifle creativity. They are meant to avoid costly inconsistency. When every team picks its own tools and patterns, support costs rise. Integrations get complicated, and risk grows. Standards create guardrails. Within those guardrails, teams can innovate safely.

What Are Reference Frameworks?

While standards lay down the rules, reference frameworks shape the big picture. They map how a company thinks about a specific area without diving into blueprints or wiring diagrams. This isn't detailed design work. It's a high-level model or a guide that steers choices before anyone starts building.

Picture a reference architecture for digital services customers use every day. You'd see clear layers: presentation, API gateway, business services, data services, shared infrastructure. Common building blocks show up too, such as identity management, logging, monitoring, messaging. What you won't find are vendor names or exact setup instructions. The goal is alignment, not configuration.

Where should business logic live? How are apps supposed to connect? Which data exchange patterns are okay, and which are off-limits? A reference framework settles those debates before implementation starts. It gives teams a shared mental map so systems fit together instead of fighting each other. Without that structure, teams often build their own version of "right". Standards define what's permitted, while reference frameworks explain how the pieces connect and work together.

Why They Matter

Standards and reference frameworks serve three main purposes: consistency, speed, and risk management.

Consistency reduces complexity. When applications use the same authentication method, support teams build deeper expertise. When data follows shared definitions, reporting becomes more reliable. Architecture is easier to understand because patterns repeat.

Speed comes naturally with these pieces in place. Teams no longer debate basic decisions for every project. Instead of asking, "Which cloud provider should we use?" they look to the approved platform standard. Instead of inventing an integration approach, they use the integration reference architecture. This leads to less debate and more delivery.

Risk management is often the hidden benefit. Security standards lower exposure to threats. Data governance standards safeguard sensitive information. Infrastructure standards make sure resilience and disaster recovery requirements are met. When these expectations are defined upfront, projects are less likely to miss them.

When standards are clear, people can see that the architecture is deliberate. The group isn't chasing every shiny new tool that pops up. Choices line up with a steady way of thinking, not with whatever trend is loudest this month.

Standards also shape culture. They tell everyone that the work is guided by purpose, not impulse. Instead of zigzagging from one technology craze to the next, the organization follows a shared blueprint. Decisions connect back to a common plan, a north star people can point to when debating a design.

Types of Standards in Enterprise Architecture

Standards cover many areas, reflecting how broad Enterprise Architecture really is.

The most visible are technical standards. These define the approved operating systems, database platforms, cloud vendors, middleware, and programming languages teams are expected to use. For instance, an organization may choose PostgreSQL and Microsoft SQL Server as its sanctioned relational databases, along with clear guidance on when each should be selected.

Integration standards define how systems talk to each other. They may require RESTful APIs with JSON payloads for real-time integration and message queues for asynchronous communication. This includes authentication protocols such as OAuth 2.0, encryption requirements like TLS 1.2 or higher, and API versioning rules.

Data standards cover naming conventions, master data definitions, metadata requirements, and retention policies. A data standard might spell out what counts as a "Customer" across systems, how customer IDs are created, and which fields are required.

Security standards set the minimum guardrails an organization must follow. They spell out rules for strong passwords, encrypting stored data, separating networks, scanning for weaknesses on a set schedule, and keeping detailed logs. Most companies base these rules on well-known frameworks like ISO 27001 or NIST, then adjust them to fit how they actually operate because no two organizations work exactly the same.

Business architecture has standards too. Teams might agree on one consistent way to describe capabilities or define KPIs so plans from different departments line up and can be compared without confusion. Without shared rules, how would anyone know if one team's metrics mean the same thing as another's?

What Goes Into a Reference Framework?

A reference framework focuses more on structure and visual guidance than a standard does. Instead of rules alone, it usually combines diagrams, layered models, and explanations that show how parts fit together.

Consider a typical application reference architecture. It may include:

- A presentation layer (web, mobile, external portals)
- An API gateway layer for routing and security
- A business services layer containing domain logic
- A data access layer
- Shared services such as logging, monitoring, and identity
- Infrastructure components (cloud hosting, networking, storage)

The framework explains what each layer does and the rules for how they interact. For example, presentation components cannot directly access the database; they must call business services through APIs. This rule keeps responsibilities separate and makes the system easier to maintain.

Another example is a data reference framework. It may distinguish between operational data stores, analytical data warehouses, and data lakes. It clarifies which systems serve as systems of record, how data moves between environments, and where data quality checks occur.

This works at the company level too. A capability framework lays out everything the business actually does and groups it into steady, long-lasting areas like Marketing, Operations, Finance, and Technology. It's a simple map of how the company runs, what it's built to be good at, and it doesn't change every time a new project comes along.

That matters because this structure becomes the backbone for other key tools. Roadmaps, budget choices, and big investment choices all lean on it. When you're clear on what the organization is capable of and where it needs to grow, you make smarter calls about where to spend time and money.

How Standards and Frameworks Are Created

These artifacts take shape through a mix of strategy, hands-on experience, and accepted industry practice. It usually starts with assessment. Enterprise Architects examine the current technology portfolio, spot duplication, and dig into performance problems. Patterns surface fast: five reporting tools doing roughly the same thing, or multiple integration methods running side by side without a clear reason. That's when the need for standards becomes obvious.

Then strategy comes into play. A company moving to the cloud will see that reflected in its standards. One focused on rapid product testing may favor modular architectures and reusable services instead because speed beats elegance when you're experimenting.

Industry frameworks provide a starting point, but they rarely fit perfectly. Architects may take existing standards, then adjust them to match the organization's size and risk level. A mid-sized company simply doesn't need the same level of formality as a global bank.

Standards pushed out without input are often ignored. Architects should bring in application teams, infrastructure engineers, security specialists, and business stakeholders early. Workshops and design reviews help check whether proposed standards are practical. This builds shared ownership. Draft standards and reference frameworks go before governance forums for a detailed review. Reviewers debate, revise, and ultimately sign off. Approval gets documented in writing, and the final materials are posted in a shared repository where anyone who needs them can find them.

Clear version numbers, straightforward change logs, and a simple update process keep everyone aligned so nobody is guessing which document is current. Updates are rolled out carefully so people aren't blindsided by surprise changes. After all, what's the point of a standard if no one knows which version to follow?

Sample Standard Document Structure

Most enterprise standards follow a familiar structure:

1. **Purpose and Scope** – A clear statement of what the standard covers and who must follow it.
2. **Policy Statement** – The required (or recommended) rules.
3. **Technical Specifications** – Concrete details such as approved versions, configurations, and protocols.
4. **Exceptions Process** – The steps teams take when they need to diverge from the standard.
5. **Compliance and Review Cycle** – How the organization

checks adherence and how often the document is updated.

Take a Cloud Hosting Standard as an example. It might require that all new applications run on the company's approved cloud provider, rely on managed database services when practical, and connect to centralized identity management. The document would also spell out the path for exceptions, perhaps asking for a written risk assessment and sign-off from an executive.

Specific language makes standards work. When a document says "Use secure encryption," teams are left to guess what that means. Stating "All data in transit must use TLS 1.2 or higher" removes that ambiguity and sets a clear bar.

Governance and Enforcement

Without governance, standards are just suggestions. Governance ensures that projects follow architectural expectations before major investment is made.

In the Kettera EA Methodology, standards are reviewed at key points in the delivery cadence. When a new initiative enters solution design, architects check alignment with existing standards and reference frameworks. If deviations are necessary, they are written up as Architecture Decision Records and formally reviewed.

Enforcement doesn't mean being rigid. There are valid reasons to deviate such as new technology capabilities, vendor limits, or unique business needs. The key is transparency. Deviations should be deliberate and time-bound.

Technology changes and company goals shift with it. Markets move, teams grow and tools get replaced. If you don't revisit the rules, they harden into red tape instead of guardrails. Why let yesterday's assumptions run today's work? You can set a yearly or twice-a-year checkup. Look at what still fits and what doesn't. Tweak, trim, or rewrite as needed, before small gaps turn into big slowdowns. Regular reviews keep standards useful instead of turning them into obstacles.

Common Pitfalls

Some organizations end up with far too many standards. The documentation grows dense, reviews take longer, and delivery starts to drag. Others swing the opposite way and define almost nothing, which leads to fragmentation and rework. The right approach is to concentrate on the areas that carry the most risk and value such as security, integration, data, and core platforms.

It's also common to copy industry frameworks word for word. While external standards can point you in the right direction, they rarely fit perfectly out of the box. Every organization has its own constraints and priorities, so adapting those frameworks to your context is necessary.

Finally, communication is often overlooked. A beautifully designed reference architecture that no one reads is useless. Training sessions, architecture portals, and tying them into project onboarding help weave these materials into daily work.

Key Takeaways

Standards and reference frameworks sit at the heart of Enterprise Architecture. Standards spell out what people across the organization must do and what they're not allowed to do. Reference frameworks lay out the big-picture structure, showing how systems and capabilities fit together.

When these tools are used well, they cut down on pointless differences between teams and keep risk in check but they don't appear by magic. Teams study the current landscape, line up decisions with business goals, work across departments, and lock things in through formal governance. Written in plain language and reviewed regularly, they guide teams without boxing them in and that balance is very important.

Next, the focus shifts from enterprise-level direction to the design of individual solutions. The standards and frameworks set the guardrails. Solution Architecture Diagrams show how those guardrails shape real systems, turning strategy into something you can actually see and build, where ideas meet code.

CHAPTER 23

Solution Architecture Diagrams

Every major initiative reaches a moment when someone leans back and asks, "What is this actually going to look like?" Spreadsheets and roadmaps won't help. A capability map or a stack of standards documents won't satisfy this need either. People want to see it. They want a visual, organized snapshot that shows how the pieces connect and where the boundaries sit. That visual is the solution architecture diagram.

In the Kettera EA Methodology, they are among the most hands-on deliverables an enterprise architect creates. They turn big-picture goals into a concrete structure people can build from. When done right, they reduce confusion before it spreads and save real money in the process.

This chapter introduces conceptual solution diagrams and explains their purpose. It describes how these diagrams are created, what they include, and how they are used in practice.

What Are Conceptual Solution Diagrams?

A conceptual solution diagram is a high-level visual view of how a proposed solution will be set up. It shows the main building blocks like applications, data stores, integration points, platforms, and sometimes external actors, and how they connect to each other. It leaves out low-level technical detail. Instead, it emphasizes clarity and shared understanding.

Think of it as the architect's sketch before construction starts. It answers key questions:

- Which big pieces make up the system?
- How do they connect and affect each other?
- Where is data created, how does it travel, and where does it live long term?
- Which parts sit inside the company, and which rely on outside vendors?
- What new abilities does this design unlock?

A conceptual diagram lives in the space between two other architecture documents. It goes further than a future-state sketch that only outlines high-level goals. But it stops short of a deep technical blueprint packed with API details, data models, server specs, and configuration settings. It stays at the altitude leaders need; clear enough to decide and not buried in wiring.

Architecture work often collapses at the extremes. Some

diagrams reduce everything to "Box A talks to Box B" which helps no one. Others flood the page with so much technical detail that stakeholders tune out. The conceptual solution diagram holds the line between those failures. It gives enough clarity to set direction while staying simple enough to walk through in fifteen minutes.

What Are Solution Architecture Diagrams Used For?

A solution architecture diagram exists mainly to create alignment. When business leaders, product owners, architects, and engineers are all looking at the same visual, confusion gets resolved. Shared understanding cuts down on rework, limits conflicting assumptions, and helps teams spot design risks early which is always cheaper than fixing them later. Teams often rely on these diagrams at the start of a project. Before approving funding, stakeholders need a clear view of the structural impact. Will the initiative introduce a new platform? Does it depend on integrating with five legacy systems? Could it lead to duplicated data? Putting the design on paper makes these effects easy to see.

Architecture review boards use them as well. With a clear conceptual view, reviewers can check whether the proposal follows established standards and reference frameworks mentioned earlier. If the company prefers a specific integration pattern, the diagram should make that visible. If cloud-first is the chosen direction, the visual should clearly show that decision reflected in the design.

Project teams rely on these diagrams as well. Developers and engineers use them to understand the broader context. A solution architecture diagram does not replace detailed design, but it sets clear boundaries. It shows where responsibilities start and stop. It defines integration expectations before any code is written.

Solution architecture diagrams also support communication beyond IT. Business stakeholders rarely want to read a twenty-page design document. They will, however, look at a one-page diagram that shows how customer data moves through a new digital channel or how a financial system connects with billing and reporting tools. A good diagram builds shared understanding without requiring technical expertise.

Key Characteristics of Effective Conceptual Diagrams

Some diagrams look amazing and still fail at their job. Others are precise down to the last detail and yet no one can make sense of them. A good solution architecture diagram lands in the middle. It communicates clearly, without showing off or drowning the reader in detail.

A conceptual diagram should stick to one initiative, one clearly defined solution space. Cramming the whole enterprise onto a single page only creates noise. The point is to make one change understandable, not to document everything the organization has ever built. If someone has to squint and ask, "What am I supposed to be looking at?" the diagram has already failed.

Major building blocks belong on the page. That usually means core applications, key data stores, the main integration paths such as APIs, messaging, or similar pipes, plus user-facing channels and any infrastructure layers that truly shape the design. Leave out the small stuff. Background jobs, minor utilities, and setup tools rarely help the reader grasp the big picture unless they genuinely alter how the system works. If they don't change the story, they don't need to be included.

Third, relationships are made explicit. Each arrow carries a label, data flows move in a clear direction, and integration types are easy to distinguish. An API call between systems should look different from a nightly batch transfer, so no one has to guess what's happening. When a diagram leaves room for interpretation, the implementation usually suffers.

Fourth, notation stays consistent throughout. There's no need to follow a universal standard, but the structure should feel familiar and predictable. External systems might sit outside a boundary box, internal applications can be grouped by domain, and data stores should share the same visual symbol. That steady use of patterns signals care and makes the diagram easier to trust.

Finally, the diagram must be readable. It sounds obvious, yet many are packed with so much detail that they overwhelm the viewer. If a stakeholder needs a ten-minute walkthrough just to grasp the basics, it's time to simplify the diagram. Clear structure and restraint make the difference.

How Conceptual Solution Diagrams Are Created

Creating a solution architecture diagram is not just a drawing exercise. It is an analytical process. The process usually begins with understanding the business objective. What problem is the initiative trying to solve? What results are expected? Without that clarity, the architecture can drift toward technology for its own sake. In the Kettera methodology, this means organizational goals drive technology decisions.

Next, identify the impacted capabilities. Refer back to the capability map from Chapter 18. Which business capabilities are improved, introduced, or replaced? This step keeps the solution grounded in the broader enterprise context rather than narrow project thinking.

The architect next figures out what pieces the system actually needs to work. For instance:

- A new web app customers can use
- A way to handle user identities and permissions
- A connection to the current CRM
- Cloud infrastructure to run everything

After that, those pieces are organized into sensible groups. Some parts will have to be built from scratch. Others are already in place and just need to be connected. Knowing what's new versus what's already there changes the budget, the timeline, and the level of risk. How can you estimate cost or spot trouble early if you don't know which parts you're actually building?

With the components defined, the architect turns to interaction patterns. How will data move through the system? Will communication happen synchronously or asynchronously? The answers shape decisions about APIs, message queues, file transfers, or shared databases. At this point, architects typically work closely with solution engineers and domain experts to test assumptions and fill in gaps.

Visual modeling comes next. Once the analysis is clear, the architect chooses a diagramming too and begins drafting the conceptual layout. Boundaries are established first, including external actors, the enterprise boundary, and any cloud environment in scope. From there, major components are arranged in a logical sequence, usually flowing left to right or top to bottom.

Arrows and labels come last. Each connection should answer a clear question: What data moves? In which direction? Under what pattern? Over-labeling adds clutter; under-labeling creates confusion. Finding the right balance takes practice.

Before finalizing the diagram, review it with stakeholders. These sessions are less about presenting and more about making sure it's right. "Is this how the process actually works?" "Are we missing a regulatory reporting feed?" "Does this create duplicate data storage?" Feedback at this stage helps avoid rework later and helps ensure alignment.

Example: Customer Portal Modernization

A company decides it's time to overhaul its customer portal. The aim is simple: let customers handle their own accounts, view and pay bills online, and open support tickets without calling anyone.

A high-level solution sketch for this effort would show:

- Customers using the system
- A brand-new web portal

- An API gateway
- The company's existing identity service
- The current CRM platform
- A legacy billing platform
- A SaaS-based support ticket tool
- A new reporting data store
- A cloud hosting setup

Customers log in and use the web portal as their front door. Every request the portal makes gets funneled through the API gateway. Sign-in and access control are managed by the existing identity provider. When someone updates their profile, that information flows to and from the CRM system.

Billing details come from the older billing platform, sometimes by way of an integration layer that bridges old tech with new. When a customer submits a support issue, the portal sends it securely to the SaaS ticketing tool through APIs. Key transaction data is then copied into the new reporting store so the business can run analytics and track trends.

Notice what's missing: server sizing, database schemas, network subnets, deployment pipelines. Those details belong in deep-dive design docs. A conceptual diagram sticks to structural relationships and nothing more.

From there, teams can put it to work. Leadership might weigh whether a new API Gateway fits enterprise standards, while security reviews identity flows and data governance examines plans for a new reporting store.

Common Pitfalls

One frequent mistake is confusing a conceptual solution diagram with a process flow diagram. While both use boxes and arrows, they answer different questions. A process flow shows the steps in a workflow. A solution architecture diagram shows system components and how they interact. Mixing the two leads to visual confusion.

Another common issue is overengineering the diagram. Architects sometimes feel the need to include every technical detail to show rigor. The result is a diagram that only another architect can understand. A conceptual artifact should communicate clearly to a broad audience. If it needs translation, it has missed the mark.

There is also the risk of treating the diagram as fixed. As projects evolve, assumptions shift. New integrations appear. Platforms get replaced. The solution architecture diagram must be updated to reflect reality. Otherwise, it becomes a historical artifact instead of a working tool.

Relationship to Other Architecture Artifacts

A solution architecture diagram lives inside a broader enterprise architecture repository, tied to the other materials that explain how the business and its systems fit together.

Each diagram points back to agreed-upon standards and reference models so the design stays in line with enterprise rules.

It also feeds updates to the application portfolio by flagging which systems are being introduced and which are on their way out. Roadmaps rely on these diagrams to make sense of timing, dependencies, and sequencing limits like what has to happen first, what must wait, and what can move in parallel. Many diagrams naturally lead to deeper, more detailed work as well, especially data relationship diagrams.

This matters because without these links, a solution diagram becomes a pretty picture instead of a working tool. Seen this way, the solution architecture diagram acts as a bridge between vision and execution. It turns strategic intent into a clear, organized design and lays the groundwork for deeper technical work.

Key Takeaways

Conceptual solution architecture diagrams provide a clear, organized view of how a specific initiative will be put into place at a high level. They show the major components, how they relate to each other, and the flow of data, without getting lost in technical details.

They are used to align stakeholders, support governance reviews, guide project teams, and spot risks early. When created carefully, based on business objectives and the impact on capabilities, they become strong communication tools, not just decorative drawings.

Building these diagrams starts with getting clear on what the business is trying to achieve. Then you pinpoint the capabilities that will feel the impact, outline the core components, and chart how they connect and interact. The finished diagram simply makes that thinking visible.

When systems lean heavily on data, a tougher question pops up: how are the data pieces actually organized and linked together? That's where a another tool comes in, data relationship diagrams. They zoom in on the structure of the information itself, exposing details that high-level solution diagrams deliberately leave out to stay simple.

CHAPTER 24

Data Relationship Diagrams

Ask most organizations what they mean by "data," and you'll hear about reports or dashboards and probably more about databases and spreadsheets. They will talk about the entities being represented but relationships between those entities help complete the picture. The real value isn't in isolated numbers or neatly labeled tables; it's in how things connect. A customer isn't just a row in a system. That customer submits orders, pays invoices, fits into a segment, and talks with support. Those links carry the story. Without them, you're staring at fragments instead of insight.

Data Relationship Diagrams (DRDs) lay those connections out in plain view. They map the ties between data entities and spell out the rules that control how those ties work. Within Enterprise Architecture, especially under the Kettera Architecture Methodology, DRDs are core tools. Teams use them to connect business goals with technical plans, and to clear up misunderstandings before anyone writes code or changes a system. That clarity saves time, money, and a lot of rework.

This chapter introduces Data Relationship Diagrams; what they are, why people use them, how to create them, and what they look like in real-world settings. You'll see how these diagrams map connections between pieces of data, along with clear examples of how they work in practice.

What Are Data Relationship Diagrams?

A Data Relationship Diagram is a visual map of data entities and the relationships between them. It shows the main objects the organization cares about such as Customer, Product, Order, Invoice, or Employee and how those objects connect.

Most readers will recognize this idea from Entity-Relationship Diagrams (ERDs), and in many cases the terms are used interchangeably. Within Enterprise Architecture, however, we usually work at a slightly higher level than detailed database design. We are less focused on column-level definitions and more focused on a clear picture: What are the core business entities? How do they relate? What are the cardinality rules between them?

At its simplest, a Data Relationship Diagram includes:

- **Entities** (usually shown as boxes)
- **Attributes** (key characteristics of each entity)
- **Relationships** (lines connecting entities)
- **Cardinality** (one-to-one, one-to-many, many-to-many)

Let's start with a simple example. A customer makes an order. That order holds one or more line items, and every line item points to a specific product. Even this tiny chain starts spelling out how the company really works. One customer can place many orders, yet each order answers to one, and only one, customer. No order exists without at least one item attached. These aren't technology details; they're core business truths.

Data Relationship Diagrams in the Kettera EA Methodology play a clear role. They connect what the business does every day with how software stores and moves that information. Think of them as the link between real-world activity and database tables.

What Are Data Relationship Diagrams Used For?

Data Relationship Diagrams have several uses, and they become more valuable as an organization becomes more complex.

To start, they build shared understanding. Business leaders, architects, analysts, and developers often use the same terms but mean different things. Is a "client" identical to a "customer"? Does an "account" refer to a billing record or a login? By putting these definitions on paper, a clear diagram brings those differences to light early on. Ambiguity doesn't stay hidden; it becomes something the team can see and address.

They also shape solution design. When a new system is being considered, architects can line up its data model against the broader enterprise view. Does the system define "Customer" in a way that conflicts with the agreed definition? Is it creating entities that already exist in another application? Asking these questions early helps prevent duplication and keeps the application portfolio in alignment.

Third, Data Relationship Diagrams support integration efforts. Integration failures are often data failures. Two systems may exchange messages successfully yet interpret the data in different ways. If one system allows multiple primary contacts per customer and another assumes there is only one, mismatches will happen. A shared understanding of relationships prevents subtle errors that surface later.

They also help with regulatory and compliance needs. Many regulations rely on understanding how data flows and connects, especially personal data. If an organization cannot clearly show how a customer's information links to orders, payments, and support cases, it will struggle to handle retention, deletion, or audit requests.

Finally, Data Relationship Diagrams help plan transformations. When an organization moves from a legacy system to a new platform, the data model has to be aligned. Knowing the current relationships, and the desired future ones, makes the transition more controlled and less risky.

How Data Relationship Diagrams Fit in the Kettera Methodology

A Data Relationship Diagram shows up when a team needs clarity about how information fits together. You'll usually see one drafted or updated during:

- Business capability analysis
- Solution design work
- Defining the current and future state
- Large system replacement or consolidation efforts

After teams build a Capability Map (Chapter 18) and spell out today's and tomorrow's operating models (Chapter 19), a pattern often appears. Different capabilities rely on the same data, yet define it in conflicting ways. Customer means one thing to Sales, something slightly different to Operations, and something else again inside Finance. That tension forces the issue. How can you redesign systems if you can't agree on what your core data actually is?

At that point, architects narrow their focus. They sketch a Data Relationship Diagram to pin down the basic entities and how they connect; the foundation everything else rests on. Used this way, the diagram serves as a reference point for conversations about who owns the data, where system boundaries lie, and how integrations should work.

How Are Data Relationship Diagrams Created?

Creating a Data Relationship Diagram is less about drawing and more about clear, disciplined thinking. The diagram is just the visible result. The real work happens in conversation.

Step 1: Identify Core Entities

Begin with the language the business actually uses. What things does the organization deal with every day? Think in terms of real-world objects, not technical artifacts. A retail operation, for example, usually revolves around Customers, Orders, Products, Suppliers, Inventory Locations, and Payments. In healthcare, the central pieces often include Patients, Providers, Appointments, Claims, and Policies.

Ignore the temptation to dive into system tables or application schemas. This isn't about dissecting a database. You're not reverse-engineering software, you're mapping how the company works. Focus on the big-picture entities that represent how the organization runs, makes decisions, and delivers value. If the business disappeared tomorrow, what core things would you need to describe how it functioned? Those are your entities.

Step 2: Define Key Attributes

Every entity is described by its attributes. A Customer, for example, might have a Customer ID, Name, Email, Date of Birth, and Status. At the Enterprise Architecture level, you don't need to document every possible field, just the identifiers and the characteristics that matter to the business. If two systems use "Customer ID" to mean different things, the diagram shouldn't smooth that over. Call out the difference so it can be resolved.

Step 3: Establish Relationships

Now connect the entities. Ask: How does this entity relate to the others? A Customer places Orders. An Order creates Invoices. An Invoice is paid with a Payment. An Employee may approve a Refund.

Define the type of each relationship. Is it one-to-one? One-to-many? Many-to-many? For example:

- One Customer can have many Orders.
- One Order belongs to exactly one Customer.
- One Product can appear in many Order Items.
- One Order Item references one Product.

This is where misunderstandings often surface. Some businesses allow joint accounts (many customers to one account). Others don't. The diagram makes that policy visible.

Step 4: Validate with Stakeholders

Resist the urge to declare the diagram finished without putting it in front of business owners and technical teams. Sit down with them and use the diagram to play out real-life situations. Have them trace what happens step by step. If a customer updates their email address, which parts of the model change? When a product is removed, what becomes of the old orders tied to it? What breaks and what should never break? Walking through these examples surfaces assumptions no one realized they were making. It also strengthens confidence in the diagram itself.

Step 5: Refine for Purpose

Not every diagram requires the same depth. An enterprise-level view might show just 10–20 core entities, while a solution-level version can drill into sub-entities and technical keys. Define the purpose and audience before you decide how much detail to include. Executives look for clarity; developers expect precision. Sometimes the best answer is two linked diagrams at different levels of abstraction.

Example: Retail Enterprise Data Relationship Diagram

Imagine a mid-sized retail company running both online and brick-and-mortar stores. A simplified Data Relationship Diagram might show the following entities:

- Customer
- Order
- Order Item
- Product
- Inventory Location
- Payment
- Shipment

Relationships might look like this:

- A Customer can create one or many Orders.
- Each Order includes one or several Order Items.
- Every Order Item points to a single Product.

- A Product can sit in multiple Inventory Locations.
- Orders can be covered by one or several Payments.
- An Order can lead to one or more Shipments.

Even this stripped-down model tells you a lot. When an order accepts more than one payment, your system has to handle split cards, gift cards, partial charges; the messy stuff real customers do. When the same product lives in different warehouses or stores, inventory tracking gets tricky. You have to manage transfers, hold stock for specific orders, and keep counts accurate across locations. Miss that detail and numbers will be off.

Now picture buying another retail company. Their setup treats "Shipment" as just a piece of the Order rather than its own separate record. That choice reshapes reporting, shipment tracking, returns, and every integration touchpoint. One structure assumes independence; the other bundles everything together, and those assumptions ripple through the entire system. The data relationship diagram turns into more than documentation. It becomes a tool during integration talks, a way to surface hidden conflicts before they blow up later.

Common Mistakes and How to Avoid Them

Overcomplication tops the list of common mistakes. Some architects pack a single page with dozens of entities and hundreds of relationships. It may look impressive at first glance, but it's useless in practice. If you need to zoom in just to read it, it won't help anyone make a decision.

Designing from the database up is another trap. When a diagram closely mirrors one system's schema, it stops reflecting the broader enterprise view. The point isn't to catalog tables; it's to make business relationships clear.

Ambiguity trips people up as well. Draw a line between two entities without stating the cardinality, and readers are left guessing. Is it one-to-many? Many-to-many? Those distinctions change how systems are built and integrated. A good diagram removes doubt rather than adding to it.

Some organizations also treat these diagrams as static documents. But data changes as the business changes. The diagram needs regular care and review, built into the architecture cadence described later in this book.

Practical Applications in Architecture Governance

In architecture review boards, Data Relationship Diagrams are often used when evaluating new initiatives. If a proposed solution introduces a new "Customer Profile" entity, the board can ask: How does this relate to the existing Customer entity? Is it an extension, a duplicate, or a specialized subtype?

When assessing integration strategies, architects can use the diagram to decide whether to integrate at the Order level, the Customer level, or both. Clear data relationships lead to clearer API contracts and cleaner system boundaries.

The diagram also helps with data ownership discussions.

Which department owns the Customer entity? Who defines the attributes of Product? Without clarity, governance becomes fragmented and accountability weakens.

Key Takeaways

Data Relationship Diagrams map how the core pieces of a business fit together. They spell out what each entity means, surface conflicts in definitions, and anchor both solution design and integration plans. They're practical tools that shape real decisions.

Building one takes discipline. You pinpoint the right entities, spell out their attributes in plain terms, lock down clear cardinality rules, and validate everything with stakeholders. Skip any of that and confusion can creep in. Done right, these diagrams cut through vagueness, steer system design across the company, and keep teams aligned.

CHAPTER 25

Technology Portfolio

Enterprise Architecture isn't just about apps and data. Under every application stack lives many technology decisions like operating systems, databases, cloud services, network gear, middleware, developer tools, security software. They add up over the years. Some were smart, planned choices. Others showed up during rushed projects, mergers, or late-night "just make it work" moments. The technology portfolio is the record that brings order to that sprawl.

This layer shows, in one organized place, the core technical pieces that keep applications and business capabilities running. Without it, you're left guessing where things are running. With it, you can clearly see what platforms you depend on and which tools the company has standardized on. It forces a basic question most leaders can't easily answer: What are we actually running, and what are the technical components holding it up?

This chapter describes what a technology portfolio is, why

organizations use it, how to build one, and what you will usually find inside. In the Kettera methodology, it serves the same purpose as other Enterprise Architecture artifacts: to guide clear decisions and real work, not to exist for the sake of documentation.

What Is a Technology Portfolio?

A technology portfolio is a structured inventory of the core technologies used across the enterprise. It lists infrastructure platforms, databases, integration tools, development frameworks, security tools, cloud services, and other shared components that applications rely on.

Unlike an application portfolio, which lists business-facing systems, the technology portfolio focuses on the underlying technical stack. Instead of listing a "Customer Relationship Management System," for example, it would name technologies such as Microsoft SQL Server, Kubernetes, AWS EC2, Azure Active Directory, Apache Kafka, or Linux.

The portfolio is usually organized by technology domain. Common categories include:

- Infrastructure (servers, storage, virtualization platforms)
- Cloud services (IaaS, PaaS, SaaS infrastructure components)
- Databases and data platforms
- Integration and messaging tools
- Application runtimes and middleware
- Development frameworks and languages

- Security technologies
- End-user computing platforms
- Networking technologies

A portfolio entry should carry more than a product name. It needs the version, the vendor, lifecycle stage, support approach, how many applications rely on it, how critical it is to the business, and who owns it. Strip that context away and you're left with just an inventory. Not a tool. Most organizations can rattle off the applications they run but how many database engines are we actually supporting? How many Java versions are live in production? Which integration platform is the real standard? Can you answer that confidently? A well-built portfolio can.

What Is a Technology Portfolio Used For?

A technology portfolio exists mainly to support governance. It gives Enterprise Architecture a factual basis for decisions instead of relying on personal preference. When a team suggests introducing a new database or messaging platform, the portfolio adds important context. Is there an approved standard already in place? Is the proposed option scheduled for retirement? How many comparable tools are we supporting today?

Another major benefit is standardization. Over time, most organizations collect overlapping technologies like three integration platforms, five reporting tools, or several container runtimes. Every extra tool adds operational overhead, increases training demands, complicates vendor management, and raises risk. By making duplication visible, the portfolio helps teams simplify and reduce unnecessary variety.

Cost management is another practical use. Licensing fees, cloud spending, hardware upkeep, and support contracts are all linked to technology choices. By grouping technologies and reviewing usage patterns, the enterprise can spot chances to consolidate or renegotiate contracts. A technology used by just one low-value application may not be worth the cost.

Risk management also relies on portfolio visibility. Unsupported software versions, end-of-life hardware, and niche tools maintained by a single engineer create real risk. The portfolio lets architects and technology leaders flag high-risk components and prioritize fixes. When a critical vulnerability is announced, the portfolio helps answer a pressing question: Where are we using this?

The technology portfolio also helps us prove that we are following a strategy. Big-picture tech slogans like "cloud-first" or "API-driven integration" sound great, but they mean little unless you can measure them. A portfolio gives you a starting line and a way to see whether you're actually moving toward that goal. It turns ambition into numbers, timelines, and visible progress. Without that grounding, strategy becomes misaligned. How do you know you're getting closer to the future you describe? The portfolio shows what you own today, what needs to change, and how far you've come. Strip that away, and the strategy stays vague.

How Is a Technology Portfolio Created?

Building a technology portfolio is less about visual polish and more about disciplined discovery. The first draft almost never gets it right. It will have gaps, inconsistencies, and rough edges, and that's fine. The aim is to create a living artifact that becomes more accurate and useful over time.

Start by defining the scope. What counts as a "technology" in your organization? Do development languages make the list? How about low-code platforms? Should SaaS infrastructure components like identity providers be included? Setting these boundaries early helps avoid circular debates later.

After that, gather the data. Common sources include configuration management databases (CMDBs), cloud provider inventories, infrastructure monitoring tools, vulnerability scanners, procurement records, and conversations with platform teams. No single source tells the whole story, so you'll need to cross-check them to confirm accuracy.

When the inventory begins to take shape, standardization becomes the next task. The same database platform may show up under different labels such as "SQL Server," "MS SQL," "MSSQL 2019." Aligning these into a consistent naming convention matters. If you skip this step, reports and analysis will quickly become more difficult to absorb.

Next comes classification. Each technology is placed in a domain and tagged with details like:

- Lifecycle status (Strategic, Standard, Tolerated, Retire)
- Vendor and support model
- Primary owner (team or individual)
- Number of apps that depend on it
- Hosting model (on-premises, cloud, hybrid)
- Version and patch level

It is also important to understand the lifecycle status. In the Kettera approach, every technology is placed into one of these buckets:

- **Strategic**: The top choice for new builds.
- **Standard**: Okay to use and fully supported, just not the first pick.
- **Tolerated**: Still allowed, but you shouldn't start anything new with it.
- **Retire**: On its way out and being removed.

This straightforward tagging system keeps architecture decisions steady across teams. When someone suggests a new tool or platform, the lifecycle label makes the direction obvious. Is it marked "Strategic"? Move ahead. If it's tagged "Retire" then it's time to rethink.

Finally, clarify the governance process. Decide who is responsible for updating the portfolio, how frequently reviews take place, and what steps are required to approve new technologies. Spell out ownership and set a clear review schedule, otherwise the portfolio will fall out of date and lead to incorrect answers.

Example: Sample Technology Portfolio Entry

Here's a sample entry for a database platform:

Technology Name: Microsoft SQL Server

Domain: Data Platform

Vendor: Microsoft

Current Versions in Use: 2016, 2019, 2022

Lifecycle Status: Standard (2019, 2022), Tolerated (2016)

Hosting Model: On-premises and Azure VM

Dependent Applications: 47

Primary Owner: Database Services Team

End-of-Support Dates: 2016 – July 2026

Notes: A migration plan is underway to move all 2016 instances to 2019 by Q4. The goal is to retire 2016 before support ends.

This amount of specificity makes it easier for people to actually do something with the information. Seeing three different versions at once makes it obvious that upgrades won't be simple. When a tool is tagged by lifecycle status, teams naturally steer new projects toward versions that are still supported. A clear migration note ties the entire portfolio back to roadmap planning, and keeps long-term strategy from drifting off course.

Picture an integration platform as another case. You could outline which systems it connects to, which versions are active, and where support deadlines are looming. That snapshot quickly shows whether the platform is stable, overdue for modernization, or quietly becoming a risk. It also helps leaders decide whether to invest more, replace it, or phase it out, decisions that are hard to make without this kind of grounded, specific detail.

Technology Name: MuleSoft Anypoint Platform

Domain: Integration

Lifecycle Status: Strategic
Deployment Model: Cloud-managed
Dependent Integrations: 112 APIs and services
Associated Standards: RESTful API design standard v3
Risks: Dependence on highly specialized developers

This kind of record gives leadership a clear view of concentration risk and the staffing demands tied to it.

Organizing the Portfolio for Insight

A static spreadsheet can serve as a starting point, but more advanced organizations move toward visual dashboards and analytics. Dashboards may show the number of technologies by lifecycle status, vendor concentration, or cloud versus on-premises distribution. Heat maps are especially useful. For example, a matrix could show technology domains against lifecycle status, highlighting areas where "Retire" technologies dominate. Another view might show application criticality against technology age, identifying systems that rely on outdated platforms.

The portfolio can also be linked with the application portfolio. This connection reveals dependency chains. A single legacy integration tool might support ten high-revenue applications. Removing it is not a simple technical decision; it becomes a coordinated program.

Some enterprises integrate the technology portfolio into their Architecture Review Board (ARB) process. Proposed projects must reference approved technologies. Exceptions require written justification. Over time, this reinforces standards and cuts sprawl.

Practical Challenges

Building and caring for a technology portfolio rarely feels exciting and keeping it accurate is tough. Tools and platforms shift constantly, especially in the cloud where services spin up and vanish in days. Automation can track a lot but people still have to review the data and confirm it makes sense.

Poor scoping of the portfolio is another trap. Add too many data fields and no one keeps them up to date. Strip it down too far and it stops being useful. The sweet spot isn't obvious. It takes judgment, trial and error, and regular cleanup to keep it practical.

Skipping the portfolio is worse. Without a clear portfolio, architecture choices turn reactive instead of deliberate. Standardization efforts stall because nobody can show what's actually in place. Leaders end up guessing about risk, cost, and redundancy instead of seeing them clearly.

Connecting to the Broader EA Practice

The technology portfolio connects directly to other core artifacts. It shapes current-state and future-state diagrams by spelling out the actual building blocks in use. It also informs roadmaps, highlighting which technologies are set for retirement and which are expected to grow. Standards and reference frameworks become more grounded because they rest on a clear, factual baseline.

Most of all, it makes technology governance measurable. Instead of debating personal preferences, architects can refer to lifecycle status, dependency counts, and support timelines. The discussion moves from opinion to evidence.

A mature technology portfolio won't remove complexity, but it does bring it into view. Once you can see it, you can decide what to change and what to leave alone.

Key Takeaways

A technology portfolio is a structured, maintained inventory of the core technologies used across the enterprise. It goes beyond a simple list by capturing lifecycle status, ownership, dependencies, and risk indicators.

Organizations rely on technology portfolios to standardize platforms, cut duplication, control costs, reduce risk, and align technical decisions with strategic goals. The portfolio serves as a practical decision-support tool for architecture governance and investment planning.

Building one requires disciplined discovery, normalization, classification, and ongoing governance. It takes effort. Its value increases over time as it connects with application portfolios, roadmaps, and review processes.

The next chapter drops the inventory and digs into the choices behind it. Architecture Decision Records spell out the thinking that shapes the tech stack like the messy debates and trade-offs that never show up on a diagram. They record why one database won out over another and why an aging platform was finally put to rest.

The portfolio tells you what's in play. ADRs tell you why it's there.

CHAPTER 26

Architecture Decision Records

Enterprise Architecture can generate a lot of artifacts: capability maps, diagrams of the current and future state, roadmaps, standards, and portfolio views. Those are the documents everyone flips through in steering meetings. What actually gives them shape, though, are the choices behind them; the decisions that pushed the design in one direction and ruled out others. Why pick this integration approach and not the alternative? Why back one platform while phasing out another? Why centralize governance instead of letting each business unit run it? Those judgments are an important part of the story.

Architecture Decision Records, or ADRs, capture that part of the story in plain terms and lock it in place. They spell out what was decided, why it was decided, and what other options were on the table. Without them, the reasoning is lost over time.

Architectural decision records are straightforward in concept. They record the key choices teams make, the situation surrounding those choices, and the trade-offs or effects that follow. Instead of stopping at what was approved, they explain the reasoning behind it. In Enterprise Architecture, where one decision can affect multiple business units, applications, and technology areas, that reasoning carries as much weight as the technical result and sometimes even more when questions arise later.

What Are Architecture Decision Records?

An Architecture Decision Record is a short, structured document that explains a single architectural decision. It usually includes the problem being addressed, the surrounding context, the options considered, the choice made, and the consequences of that choice. Each ADR covers one topic only. It is not a design document or a policy manual. It just records intent and reasoning.

ADRs began in software engineering communities as lightweight text files stored alongside source code. Over time, the idea spread beyond development teams into broader architectural practices. In Enterprise Architecture, ADRs act as institutional memory. They capture the thinking behind standards, platform choices, integration patterns, data governance approaches, and even organizational structure decisions.

What sets ADRs apart from other artifacts in this section of the book is their narrative quality. A capability map shows what the business can do. A technology portfolio shows what systems are in place. An ADR explains why those systems were chosen. It creates a clear, traceable line of reasoning from business strategy to architectural outcomes. With ADRs architecture becomes understandable, even years later, after the original architects have moved on.

What Are Architecture Decision Records Used For?

Imagine that a new CIO walks in and asks why the company picked one cloud provider over another. With an ADR, the logic is right there: cost comparisons, compliance demands, integration limits, risk trade-offs. The debate stops being about defending old calls and starts focusing on a better question: have the underlying facts changed?

Big companies can debate decisions in a cyclical nature. The same architecture discussions pop up year after year. Do we build it or buy it? Should APIs run sync or async? Do we centralize master data or let domains own it? An enterprise ADR captures the earlier thinking. Teams can still debate but now they have the benefit of the previous analysis.

Enterprise Architecture lives between business goals and the tech that makes them real. A single choice can ripple across business workflows, data models, integration patterns, and platforms. An ADR lays out the "why" so every group sees the full picture. It becomes the common reference when everyone asks why they took one path over another.

Finally, ADRs strengthen governance practices. Architecture review boards regularly approve or reject proposals, but without a written record, those approvals can become unclear over time. ADRs provide a clear history of decisions, showing which ones are active, which have been replaced, and which are still being discussed, which saves a surprising amount of back-and-forth later. Instead of relying on memory or scattered conversations, teams have a structured record that shows exactly how and why decisions were made.

How Are Architecture Decision Records Created?

Creating an ADR should not become a bureaucratic task. If writing one takes weeks, it won't be kept up. The Kettera EA Methodology emphasizes kinetic movement, decisions made with clarity and momentum. ADRs need to support that pace.

An ADR is typically created when a decision meets at least one of these criteria: it has enterprise-wide impact, it sets or changes a standard, it carries significant cost or risk, or it affects multiple domains. Minor project-level design choices do not require enterprise ADRs. Discipline in scope will prevent clutter later.

The process usually starts with a proposal. A team identifies a problem or opportunity. They document the context such as business drivers, constraints (remember the Three C's from earlier in the book: capability, capacity, and capital), existing systems, and timing. Options are outlined, not in exhaustive detail but with enough depth to show thoughtful comparison.

Then state the decision in plain, direct language. For example: "We will standardize on Platform X as the enterprise integration platform for all new initiatives."

With the decision documented, you must also document the impact. What changes now? Which systems get shut down? Who needs training, and on what? What risks are still hanging around? Readers should see the ripple effects in concrete terms.

After drafting, the ADR moves through the appropriate governance path, typically an architecture review board, for evaluation. Once it clears that step, it's assigned a status such as "Accepted," "Proposed," "Deprecated," or "Superseded." Because decisions change over time, careful version tracking is a must. When a new ADR replaces an earlier one, that link should be clearly documented so no one has to piece the story together later.

Where you store ADRs is just as important. They need to be easy to find and simple to search. Some teams keep them in shared document repositories, while others rely on collaborative tools. What matters is that anyone who needs an ADR can locate it quickly.

A Practical Structure for ADRs

Formats vary, but a consistent template improves readability and adoption. Below is a commonly used structure adapted for enterprise architecture:

Title: Clear, concise statement of the decision

Status: Proposed / Accepted / Deprecated / Superseded

Date: When the decision was finalized

Context: Background, business drivers, constraints, current state

Decision: The chosen approach

Options Considered: Alternatives and a brief rationale

Consequences: Impacts, trade-offs, and required actions

This structure keeps the record focused and easy to scan. It avoids extra narrative while ensuring nothing important is left out.

Example 1: Integration Platform Standardization

Title: Move to MuleSoft as the Company's Integration Platform

Status: Approved

Date: March 15, 2025

Context:

Three different integration tools are running across the company today. Each business unit picked its own solution, and now we're paying for it. License fees are higher than they should be. APIs built in one area rarely get reused in another. Teams have grown comfortable with their own tools, which has created knowledge silos and slowed collaboration.

New digital programs are on the way, and they need strong API management plus support for both cloud and on-prem systems. The current mix of tools struggles to scale cleanly. Compliance rules also require clear oversight, detailed audit trails, and centralized monitoring; something that's hard to guarantee when systems are scattered across platforms.

Consolidating on MuleSoft would simplify operations, reduce overlap, and make it easier to share skills and components across teams.

Options Considered:

1. Maintain existing tools with improved governance
2. Standardize on MuleSoft
3. Standardize on Azure integration tools

Option 1 would minimize short-term disruption, but it would leave the current fragmentation in place. Option 3 fits well with our Microsoft investments; however, at the time of evaluation, its API management capabilities were less developed than MuleSoft's.

Decision:

MuleSoft will become the enterprise integration platform for all new development efforts. A 24-month roadmap will guide the migration of existing integrations to the new standard.

Consequences:

Integration teams will need structured training. Legacy tools will be phased out according to a defined retirement plan. Licensing costs will rise in the short term, though consolidation is expected to reduce overall spending over time. API governance will move to a centralized model which will be a significant shift for teams used to operating independently.

Example 2: Data Governance Model

Title: Move to a Federated Data Governance Approach

Status: Accepted

Context:

Different business units run with a fair amount of independence and face their own regulatory requirements. When we tried to control everything centrally, progress slowed to a crawl and analytics projects stacked up waiting for approvals. Yet when we swung to the other extreme and let each team govern its own data, definitions drifted apart and reports started conflicting. The result was confusion, rework, and a lot of second-guessing.

Decision:

Use a federated setup. Company-wide data standards will be set at the enterprise level, giving everyone a shared foundation. At the same time, each business unit keeps responsibility for managing and caring for its own domain data, close to the source, where context actually lives. This strikes a balance between consistency and flexibility and avoids the challenges of going fully central or fully local.

Consequences:

A Data Governance Council was established. Each domain now has clearly defined stewardship roles, and a shared data glossary is mandatory. Coordination overhead has gone up, but teams respond faster and with fewer misunderstandings.

This shows ADRs aren't just for technical choices. They can also capture structural and organizational architecture decisions, not just code and infrastructure.

Common Pitfalls and How to Avoid Them

Some organizations try to use ADRs as replacements for detailed design documents. That's a mistake. An ADR should summarize the reasoning, not repeat technical specs. Keep it focused on the decision logic.

Another common pitfall is recording decisions after implementation has already happened. Retrospective ADRs can help, but they often miss the depth of thinking present at decision time. Capture decisions as they happen.

There is also a tendency to write overly defensive ADRs; documents that justify a decision instead of clearly explaining it. Be honest about trade-offs. If cost mattered more than technical advantage, say so.

Finally, avoid overproduction. Not every architecture discussion needs an ADR. If everything is documented, nothing stands out. Reserve ADRs for decisions that truly shape the enterprise.

ADRs Within the Kettera EA Methodology

The Kettera method treats Enterprise Architecture as something practical: make things clear, keep things moving, show real results. ADRs help on every front. They spell out why a choice was made instead of leaving people to guess. They cut off the endless rehashing of old arguments. They also create a visible link between big-picture strategy and the systems that eventually get built.

ADRs tie everything together. A roadmap (Chapter 20) can point to specific ADRs that explain why certain initiatives come first and others wait. A technology portfolio (Chapter 25) often mirrors the standards and boundaries laid out in those decisions. Capability maps shift over time as structural calls, captured in ADRs, reshape priorities. Think of these records as the stitching that holds separate diagrams and plans in one piece.

This is where maturity shows. When teams consistently write decisions down, architecture stops revolving around whoever speaks the loudest, or has been around the longest, and starts resting on shared principles. New architects can read past ADRs and quickly grasp how and why the organization made its key moves.

Key Takeaways

Architecture Decision Records (ADRs) capture important architecture choices, laying out the context, the alternatives considered, and the consequences that follow. By recording this reasoning, teams avoid rehashing the same debates months later and keep hard-won knowledge from disappearing. They also help business and technical groups stay in sync. When teams create ADRs regularly and review them through governance processes, decisions are easier to track and ownership is clearer.

A strong ADR sticks to one decision and explains the trade-offs plainly. Write it when the decision is made, store it where everyone can find it, and keep its status up to date. While not every small choice needs an ADR, decisions that shape the direction of the enterprise should be written down.

As we move into the next section of the book, we shift from artifacts to cadence. Artifacts such as ADRs provide structure and clarity, but they do not stand alone. They need to live within a steady rhythm of review, refinement, and action. Chapter 27 explores Architecture Cadence and highlights the key difference that turns static documentation into real, sustained enterprise movement.

Enterprise Architecture Cadence

CHAPTER 27

Architecture Cadence - the Kinetic Difference

Architecture cadence is the engine that turns documents into movement. It creates momentum and keeps things from stalling. Think of it as the steady cycle of reviews, decisions, working sessions, and real deliverables that keep Enterprise Architecture woven into everyday business. This isn't about booking recurring meetings or adding red tape, it's about building a repeatable rhythm that turns good intentions into work that actually gets done.

In the Kettera Enterprise Architecture Methodology, cadence provides the momentum. It keeps capability maps current, refines roadmaps over time, ensures standards are actually put into practice, and builds a clear trail of decision records. Without a steady rhythm, architecture starts to wander but with it, the work gains direction and pace.

From Static Design to Continuous Motion

Most organizations do not fail at defining architecture. They fail to keep it going.

An enterprise architect may produce a current-state diagram and a future-state roadmap. For a short time, everyone is aligned. Then projects speed up, priorities shift, and new ideas appear. Six months later, the architecture documentation no longer reflects reality. The roadmap gathers dust. The standards are applied unevenly and the architecture team ends up reacting instead of leading.

A good cadence keeps things aligned. It creates a predictable cycle of review and delivery. For example, a quarterly architecture review may recheck capability maturity and confirm the 12–24 month roadmap. A monthly standards review may update technology guidance based on active projects. A biweekly solution review board may review new initiatives against reference frameworks and architectural constraints.

Think of it as the operating system for architecture governance, the layer that makes everything else function. Strip it away and even the smartest framework turns into theory, nice on paper and useless in practice.

The Components of Effective Architecture Cadence

An effective architecture cadence rests on three defining elements: frequency, artifact production, and decision integration.

Frequency creates predictability. When stakeholders know architecture reviews happen monthly or quarterly, they plan around them. Project timelines line up with governance checkpoints, and technology evaluations are ready ahead of time. Instead of feeling like an interruption, architecture becomes part of the regular flow of work.

Artifact production ensures that each cycle produces something concrete. A cadence that yields no updated artifacts is merely performative. Each session should lead to visible change: an updated capability map that reflects new strategic priorities, a revised application portfolio showing system retirements, or an architecture decision record that captures a newly adopted integration standard.

Decision integration ties architecture to business execution. If architecture meetings do not shape funding, prioritization, or solution design, they become advisory at best and ignored at worst. The cadence must connect directly to project approvals, budget planning, and technology choices.

Consider a practical example. A company launching a new digital customer channel might trigger several architecture touchpoints: a solution architecture review, a data impact assessment, and a standards compliance check. With a defined cadence, these reviews happen at predictable intervals, producing updated diagrams and decision records. The roadmap shifts to reflect required platform upgrades. Resource planning adds the newly identified integration work. The organization moves together.

Without a proper cadence, each of those activities would occur inconsistently, if at all.

Cadence as Organizational Discipline

In many organizations, architecture is viewed as either a bottleneck or something optional. When engagement with EA depends on personal connections or one-off requests, uneven results are inevitable. Some projects follow every guideline; others skip architecture altogether. Standards end up meaning different things in different departments.

A clear cadence removes that gray area. It makes it explicit that initiatives above a defined threshold will go through architecture review at set checkpoints. It clarifies when roadmaps are updated and when standards are revisited. Over time, this rhythm becomes embedded in how the organization operates. This doesn't require long, bureaucratic sessions. Shorter, more frequent reviews often work better than infrequent deep dives. A two-hour monthly standards meeting can accomplish more than a single annual overhaul. With a steady rhythm in place, issues stay small and easier to address, and far less dramatic when they surface.

There's also a psychological side to this. Momentum builds confidence. When teams see architecture consistently shaping project direction by approving technologies, clarifying integration patterns and phasing out redundant systems, they start to see its value. Architecture stops feeling like abstract theory and becomes real.

Producing Tangible Results

A steady cadence only counts when people across the company can see what it delivers. In the Kettera approach, progress shows up as clear, usable outputs like capability maps, current and future state diagrams, roadmaps, and application portfolios, along with the standards and decision records that support them. These are not one-time documents. They live, change, and age. Each one needs a clear owner and a regular refresh cycle, otherwise it becomes outdated and ignored.

A roadmap that spans 12 to 24 months cannot sit untouched for the full stretch. Markets move, business goals evolve and new regulations appear with little warning. Reviewing the roadmap every quarter keeps it grounded in what's happening now while still aiming at the long-term destination. Small course corrections replace massive resets. The direction stays steady, even as the exact path shifts.

Application portfolios need the same discipline. Twice a year, step back and examine what's running. Which systems overlap? Which are too expensive? Which no longer earn their keep? Some can be merged. Others should be upgraded or shut down entirely. Those decisions flow straight into budgeting and staffing plans. Over time, this steady trimming lowers costs and cuts down technical debt without dramatic, risky upheaval.

Architecture decision records (ADRs) are another example. When teams capture them consistently through regular, cadence-driven reviews, they create a running history of why specific technologies or patterns were selected. That record keeps the same debates from resurfacing and helps teams move faster the next time a similar choice comes up. Over time, the collection becomes more useful as it grows.

The benefits build gradually. One review might not seem like much, but five years of steady cadence can shape a mature, coherent architecture.

Aligning Cadence with Business Rhythm

Architecture can't run on a timeline that's disconnected from the business. If the organization conducts annual strategic planning every October, architecture roadmap updates should happen before or align with that cycle. If budgeting occurs quarterly, portfolio reviews need to feed into those decisions. If product launches follow a predictable release calendar, solution reviews should tie into those milestones.

When architecture cadence mirrors business cadence, it feels natural instead of imposed. Architecture becomes part of decision-making rather than a late-stage reviewer.

Imagine a retail organization preparing for a seasonal surge. Technology upgrades, integration improvements, and performance testing must line up with that seasonal window. A cadence that schedules roadmap adjustments in the off-season gives teams time to prepare. A misaligned cadence creates friction and rework.

Architecture isn't the metronome of the business but it has to keep the same rhythm. The company sets the pace; architecture follows it and adapts when it changes. If the two fall out of sync, progress stalls. When they move together, everything feels almost effortless.

Avoiding the Two Extremes

Most organizations gravitate toward one of two poles: too much governance or almost none.

When governance goes overboard, it shows up as endless reviews, stacked committees, and documentation that grows faster than delivery. Work slows to a crawl, frustration builds, and architecture starts to look like a blocker instead of an enabler. In that setting, the steady rhythm of planning and review turns into background noise with more motion than progress.

At the other end is laissez-faire architecture. Reviews become optional, standards feel negotiable, and roadmaps read like wish lists. Teams may move faster in the short term, but problems soon appear. Duplicate systems multiply, integrations tangle, and costs creep upward until they're impossible to ignore.

The real difference comes down to balance. A clear cadence provides enough structure to guide decisions without slowing down execution. It makes clear when architecture involvement is needed and when teams can move forward on their own. It sets expectations without micromanaging. This balance is not achieved once and kept forever. It must adapt. Which brings us to the next logical step.

Measuring the Health of Cadence

Is your architecture cadence actually doing its job?

Proof shows up in the work itself. Check the artifacts and the results they drive. Capability maps should be up to date and people should actually pull them into strategy discussions. Roadmaps ought to surface in budget debates, shaping where money goes instead of collecting dust. Solution teams should book architecture reviews on their own, without nudges or calendar policing. Standards need to show up in real projects, not just in slide decks.

Success can be judged by behavior. When leaders argue about priorities, do they reach for the architecture materials? When funding gets tight, do the roadmaps guide trade-offs? Teams that value the cadence don't treat it like a compliance chore, they treat it like a tool. If you have to chase people down, something's off.

Another sign is less rework. When architects are involved early and stay involved, projects are far less likely to need major redesign once build work is underway. Integration problems show up earlier, and technology conflicts get settled before contracts are locked in.

Are business leaders asking for architectural input when they size up new opportunities? Do technology teams refer to standards on their own, without being nudged? When architecture becomes part of everyday decision-making and the normal rhythm of how choices get made, the cadence is working.

Trouble shows up in small ways at first. Artifacts go stale. Reviews get pushed back again and again. People navigate around the process. That usually means the timing is off, the scope is too wide, or no one is clear on what's expected.

Key Takeaways

Architecture cadence is the driving force that turns Enterprise Architecture from static documentation into steady forward movement. It creates predictable rhythms for reviews, artifact creation, and decision integration. By setting a clear frequency, producing tangible outputs, and aligning with business cycles, cadence weaves architecture into everyday operations.

An effective cadence delivers visible results: updated roadmaps, refined portfolios, documented decisions, and stronger alignment between strategy and execution. It blends discipline with flexibility, avoiding both red tape and unmanaged autonomy. Over time, this steady momentum builds real organizational maturity.

Cadence alone, however, is not enough. Structure brings stability, but too much rigidity can limit adaptation. The next chapter explores how to design a practical architectural structure that still allows flexibility so cadence fuels progress without holding back innovation.

CHAPTER 28

Concrete Structure with Adaptability

Enterprise Architecture cadence was described earlier as the kinetic edge or the steady beat that pushes static analysis into real change. But a rhythm in theory is useless. It has to show up somewhere. It needs calendar invitations, clear decision gates, written outputs, and people whose names are actually attached to the work. Without those anchors, cadence is just a nice idea but if you add too many rules, and it becomes more of a burden than a benefit.

In this chapter you'll find a working model that shows how Enterprise Architecture functions day to day. Expect defined activities, visible handoffs, and moments where real decisions are made. You'll also see how an idea travels from a rough concept to something running in production, through conversations, trade-offs, and approvals. However, none of this is meant to be prescriptive. The point is impact, not ideology. The framework exists to produce results, not to win arguments about process purity.

At the center of the Kettera approach is a straightforward operating cycle: **Assess, Anticipate, Actualize**. That rhythm does not change. What does change are the touchpoints, the people involved, and the artifacts produced, all shaped by the organization's size, culture, and comfort with formality. The methodology itself is concrete, but how it plays out in practice can and should adapt.

The Assess, Anticipate, Actualize Cycle

Every healthy Enterprise Architecture function works in a loop.

Assess means understanding the current state. What capabilities are in place? What systems are being used? Where are the pain points? What commitments are already funded? Assessment isn't an annual exercise. It's ongoing. It comes from portfolio reviews, performance metrics, and regular conversations with business leaders.

Anticipate looks ahead. What strategic objectives are taking shape? What market conditions are changing? Which technologies will impact the organization in the next twelve to twenty-four months? This is where architecture shifts from documentation to real insight. Anticipation calls for pattern recognition and focused imagination.

Actualize is where plans hit the ground. Big ideas turn into real programs. Roadmaps secure money and owners. Standards stop being slides and start steering design calls. Architecture shows up in solution sketches, sprint priorities, delivery guardrails, and the hard limits engineers build around.

But how does this look in practice?

Concrete Activities That Drive Results

Take a mid-sized organization with a formal IT department and clearly defined business units. In this setup, the Enterprise Architecture rhythm could include:

- A monthly portfolio review with business and technology leaders
- A quarterly roadmap alignment session
- Biweekly architecture review boards for new initiatives
- Ongoing upkeep of architecture decision records
- A rolling 12–24 month technology roadmap

These are not theoretical ideals. They are calendar invitations with names attached. They produce real outputs: updated roadmaps, documented decisions, clarified priorities, and sometimes hard "no" decisions.

In a small company, this cycle shows up in humbler ways. The "architecture review board" might just be a 45-minute working session with the CTO and two senior engineers around a whiteboard. Roadmap alignment for the quarter could happen inside a larger executive planning meeting. And instead of thick binders or polished repositories, notes may live in shared docs that everyone edits on the fly.

Formality isn't the point. The real question is simpler: are you consistently Assessing, Anticipating, and Actualizing in a way that repeats and produces results you can actually see? If those three actions, done again and again drive steady, measurable gains, the structure around them can be as lightweight or as polished as you like.

When Enterprise Architecture turns into pure ceremony, people stop taking it seriously. Yet when it operates without any structure at all, it stops having real impact. The balance sits somewhere in between, structured enough to matter, flexible enough to stay relevant.

The Flow from Idea to Implementation

A central purpose of cadence is to give ideas a clear path from concept to delivery. Without a defined flow, innovation stalls or skips architecture entirely.

In the Kettera methodology, the flow usually follows this sequence:

1. Idea generation
2. Initial screening and refinement
3. Value and impact analysis
4. Prioritization against other initiatives
5. Resource planning
6. Project approval
7. Transition into delivery

These steps are covered in detail in later chapters. Here, we focus on how they link to architecture cadence.

A marketing leader brings forward an idea such as, build a new customer analytics capability. The proposal lands in the pipeline during an Assess conversation. Teams capture what current reports can't do. They look closely at the systems already in place. They also map out key dependencies and constraints. In this way, the gaps become visible, not just assumed.

Architects then examine how the idea fits the bigger picture. Does it drive revenue? Will it demand new data platforms? Could regulations create risk? Will current teams even have the skills to support it? Future-state diagrams and roadmaps help answer those questions and they make the abstract concrete.

The idea then turns into a funded initiative with real backing. Solution architects draft detailed designs with standards guiding the way. Technology options are checked against what the company already owns and supports while the effort plugs into the organization's delivery engine.

The concept didn't jump from excitement straight into build mode. It also didn't get buried in endless debate. Finding the balance between action and analysis is one of the more difficult aspects of running an architecture practice and the right approach will vary across organizations.

Adjusting the Level of Rigor

Not every organization needs the same level of rigor. A global enterprise under regulatory oversight requires more documentation and formal checkpoints than a regional startup working in a low-risk setting.

In highly regulated industries, Assess might include formal audits and compliance reviews. Anticipate could mean risk committees and scenario planning sessions. Actualize may be held up by strict approval boards and clear traceability to standards. In this kind of setting, the architecture team has to be comfortable with disciplined processes.

On the other hand, in a rapidly scaling tech company, moving quickly usually beats following ceremony for its own sake. "Assess" often happens through real-time dashboards and quick, candid chats instead of polished reports. "Anticipate" may unfold at an offsite strategy session rather than inside a thick planning binder. "Actualize" is typically woven straight into agile sprints, guided by simple architectural guardrails. Just enough structure is provided to keep things from falling apart.

It only works if the loop stays complete. What happens when one step drops out? Skip Assess and you miss risks. Ignore Anticipate and technical debt accumulates until it slows everything down. Leave Actualize without architectural guidance and systems start to splinter. Break the rhythm and the process will break down.

Roles and Touchpoints

Another area where flexibility matters is participation. In some organizations, Enterprise Architecture is centralized, with a Chief

Architect overseeing domain architects. In others, architectural responsibilities are spread across senior technologists.

The Assess phase may involve business unit leaders in one company and mostly technical stakeholders in another. Anticipate sessions might include strategy teams, finance partners, or product managers, depending on the structure. Actualize almost always includes delivery teams, but the way they engage varies.

Instead of prescribing a fixed list of attendees, the Kettera methodology focuses on clear accountability. Who owns the roadmap? Who approves deviations from standards? Who makes sure architectural decision records are captured? If a new idea shows up tomorrow, where does it land? Any company should be able to point to a first stop without hesitation.

Results Over Dogma

Architecture groups love to argue about frameworks, modeling languages, and governance models. Sometimes those discussions sharpen thinking. Other times, they drain energy from the work that actually matters. The goal isn't strict compliance with a method. It's better performance across the organization.

Ask simpler questions. Are redundant systems being reduced? Are decisions made earlier so delivery moves faster? Are investments clearly tied to strategic priorities? Those are the measures that count. If a ceremony or artifact doesn't help move those numbers, it's worth asking why it exists at all.

Rigid process breeds frustration. Business leaders can tell

when architecture is defending its own rituals instead of helping the company move forward. But when architects show flexibility, without sacrificing sound structure, they're invited into real decision-making conversations.

Embedding the Cycle into Culture

The Assess–Anticipate–Actualize cycle needs to seep into the way the organization thinks and acts. Leaders start asking what's next without being nudged. Project pitches point to roadmaps on their own. Teams write architecture decision records because they see the upside, not because someone told them to.

Real traction comes from visible wins with clear, undeniable proof. A predicted tech shift avoids a lot of rework, and people pay attention. An early review stops two groups from building the same big system, and trust grows. Over time, cadence stops feeling like something imposed from above and starts to feel shared. People come to expect that ideas will be reviewed in a structured way, that long-term effects will be weighed, and that delivery will follow agreed standards.

By then, Enterprise Architecture isn't an added layer sitting on top of the work. It moves with the day-to-day flow of operations, woven into how decisions actually get made.

Key Takeaways

Concrete structure is essential if Enterprise Architecture is to deliver measurable results. Named activities, clear touchpoints, and documented outputs create clarity and forward motion. Without them, cadence slips into informal conversations that rarely lead to execution.

At the same time, no two organizations need the same level of rigor. The Assess–Anticipate–Actualize cycle stays constant, but the depth of documentation, meeting frequency, and governance formality should match the organization's context.

Above all, results over dogma must guide decisions. Processes exist to improve performance, not to satisfy theory. When structure and flexibility are in balance, Enterprise Architecture becomes disciplined and adaptable.

The next chapter shows exactly how they enter the system through what we call the Innovation Pipeline. You'll see how a rough thought moves step by step into a clear, organized review process. The path from spark to scrutiny becomes visible and the system's rhythm starts to feel concrete. This is the first obvious doorway into the steady cadence we've been describing.

CHAPTER 29

Innovation Pipeline

Ideas can come from anywhere but turning the right ones into real change is the hard part. In most companies, suggestions pop up in hallway chats, planning sessions and moments of day-to-day frustration. Few make it past the discussion though. When there's no clear structure to carry the idea to action, progress will stall. When no one owns decisions, efforts will flounder and when ideas ignore the bigger system, the result is a patchwork of disconnected fixes.

The Innovation Pipeline creates a clear route from spark to rollout. Inside the Kettera Enterprise Architecture approach, this process keeps innovation tied to strategy, resources, and architecture so that energy turns into results.

This chapter walks through the pipeline from start to finish. Later chapters zoom in on each stage and explore how it works in practice. For now, the goal is to see the whole system at once and to understand how Enterprise Architecture holds it all together.

From Idea to Implementation

The Innovation Pipeline includes seven linked stages:

Phase	Output	Decision Gate	Key Artifacts
Idea Generation	Raw Ideas	Idea logged with basic description and sponsor	Idea Intake Form, Idea Backlog
Idea Refinement	Feasible Ideas	Feasibility confirmed (technical + operational)	Initial Solution Outline, Feasibility Notes
Value Analysis	Valuable Ideas	Business value defined and measurable	Business Case, Value Assessment, High-Level Estimates
Prioritization	Most Important Ideas	Priority approved against strategic objectives	Prioritization Matrix, Scoring Model, Ranked Backlog
Resource Planning	Resource Plan and Roadmap	Resources allocated and timeline validated	Roadmap, Resource Plan, Dependency Map
Approval	Project Charter	Formal approval and funding authorized	Project Charter, Funding Approval, Governance Record
Transition	Active Implementation	Delivery team aligned and execution initiated	Handoff Document, Delivery Plan

Each stage answers a different question. Is the idea worth discussing? Is it clear enough to evaluate? Does it create measurable value? Does it fit our strategy? Can we afford it? Should we commit? How do we ensure it fits with what already exists?

This flow builds momentum without pushing anyone to commit too soon. Early on, an idea just needs to show promise. As it moves ahead, the scrutiny tightens and each step raises the bar. Think of the pipeline as a funnel, wide at the top, strict at the bottom, with only the strongest concepts making it through.

Enterprise Architecture stays involved across all seven stages. Architects don't just show up at the final approval gate. They work alongside teams to shape rough ideas, identify what will change, surface hidden dependencies, and keep the overall design clean and consistent from start to finish.

Why a Pipeline Is Necessary

Organizations tend to swing to one of two extremes. In some, every idea turns into a project, leading to overload and scattered efforts. In others, only a small executive group can introduce change, which slows progress and discourages new thinking. An Innovation Pipeline creates a middle ground, pairing openness with clear guardrails.

Take a mid-sized financial services firm that receives dozens of proposals each quarter such as new analytics tools, customer apps, automation efforts, compliance updates, and vendor platform changes. Without a clear process to refine, prioritize and plan initiatives, leaders compete for budget and visibility based on urgency or political pull. Technology teams end up reacting instead of planning. Over time, disconnected systems add up and the company loses sight of its strategy.

Now imagine that same firm using a defined pipeline. Every proposal enters through a shared intake process, and ideas are documented in a consistent way. Teams review architectural impact early and flag duplicate efforts before money is committed. Leaders weigh options against agreed-upon criteria instead of personal excitement. The discussion shifts from "Who is shouting the loudest?" to "Which initiative moves us toward our stated goals?"

The Expanding and Narrowing Flow

The pipeline starts wide and becomes progressively more selective. During Idea Generation, the goal is volume and diversity. Suggestions may come from frontline employees, architects, executives, or customers. At this stage, judgment is minimal. The aim is simply to surface opportunity.

As ideas move into Refinement, they are clarified and scope is defined while assumptions are documented. Many ideas naturally stall here, not because they are rejected, but because they lack clarity or ownership. This is actually fine since the pipeline filters ideas without confrontation. If the idea is too ambiguous or doesn't attract an owner then it would likely fail later, anyways.

Value Analysis introduces measurable evaluation. Financial impact, operational efficiency, risk reduction, compliance needs, and customer outcomes are all considered. Architects help by mapping each idea to capability gaps, application impacts, and data implications. An idea that sounded compelling in a meeting may reveal hidden integration complexity or duplicate existing tools at this stage.

Prioritization forces comparison across the candidate ideas. Not every valuable initiative can move forward at once. Trade-offs become visible in this stage. Does the organization pursue a revenue-generating enhancement or a risk mitigation effort? Does it modernize infrastructure now or wait until next fiscal year? Enterprise Architecture provides the context needed for informed trade-offs through roadmaps, capability maps and portfolio views.

By the time a proposal gets to Resource Planning and Project Approval, it has already been put through its paces. People have poked at it, questioned it, stress-tested it from different sides. What's left is a sturdier idea that is ready to become reality with fewer blind spots once work begins.

The Architect's Role Across the Pipeline

Enterprise Architects don't always "own" the pipeline. That accountability usually rests with executive leadership or portfolio management. Still, architects shape what happens at every step.

At the front end, they turn loosely formed ideas into clear architectural direction. A marketing leader might ask for "a better customer personalization experience." The architect translates that ambition into specific elements such as data integration, customer master records, analytics platforms, and the applications where those insights will appear. What starts as a broad goal becomes a set of defined building blocks.

During Value Analysis, architects spot downstream effects. A reporting tool that seems isolated might require changes to core transaction systems. A new SaaS platform could duplicate identity management capabilities already in place. By surfacing these issues early, architects prevent costly rework later.

At Prioritization, architecture roadmaps provide important context. If the enterprise has already committed to consolidating CRM platforms, a new standalone CRM proposal deserves close scrutiny. Without architectural visibility, leadership may unknowingly fund initiatives that clash with the approved direction.

Finally, during Transition, architects make sure approved projects align with standards and fit into existing environments. The pipeline doesn't stop at approval. It carries through disciplined execution.

Governance Without Gridlock

People often worry that formal pipelines choke off creativity and that can happen. When governance is built poorly, it turns into red tape with too many layers of review, stacks of paperwork and no final decision. An effective approach tackles this head-on by setting clear decision thresholds. It spells out who decides what, and when, so ideas don't get stuck waiting for permission.

Not every proposal deserves the same level of inspection. A small tweak to a tool already in place can move ahead fast with a light check. Replacing a core platform with a multi-million-dollar investment? That calls for serious digging, risk evaluation, and more input. The pipeline adjusts its level of scrutiny based on potential impact. Bigger bets get deeper review. Smaller changes move quickly. That balance keeps control in place without grinding progress to a halt.

People should be able to see where their idea stands in the pipeline and understand why it's moving ahead or getting stuck. Keep the backlog visible, write down the evaluation criteria, and record decisions so rejection never feels random. Architects help by explaining trade-offs and technical implications in plain language instead of jargon.

Connecting Strategy to Execution

The Innovation Pipeline connects strategy to delivery. Strategy sets the direction and projects create results. The pipeline makes sure what gets built reflects what leadership asked for.

Imagine an organization that declares digital self-service a strategic priority. Without a pipeline tied to that strategy, funding might still go to unrelated internal tools or isolated department requests. With a clear review process, each proposal is matched against strategic themes. If digital self-service ranks high, initiatives that directly support it receive higher priority scores.

Enterprise Architecture strengthens this alignment by linking initiatives to capability maps and future-state plans. When an idea enters the pipeline, architects can show where it fits in the broader transformation roadmap. Is it moving the organization toward a defined future state? Is it a temporary fix? Is it redundant? This visibility creates discipline. Leaders see not only individual project benefits but the cumulative impact of those projects.

Managing Risk and Capacity

The pipeline also guards organizational capacity from over allocation. Each approved initiative draws on budget, skilled people, and executive focus. When leaders lack a clear view across the portfolio, overcommitment follows. Teams end up juggling too much, deadlines slide, and morale takes a hit.

Resource planning inside the pipeline injects some realism. Do we actually have the architects, developers, analysts, and operations staff to deliver? What work is already underway? Could we stage delivery to ease the pressure and avoid burning out the same high performers again?

A Living System, Not a One-Time Design

The Innovation Pipeline itself needs regular review. Metrics offer clear feedback such as how long it takes for an idea to move from submission to decision or what percentage of approved projects actually deliver the expected value? Where do ideas tend to stall?

If the front end becomes overloaded, the intake criteria may need to be tightened. If too many low-value initiatives get approved, the evaluation standards may be too loose. Enterprise Architecture helps by reviewing portfolio trends and the architectural debt created by earlier decisions.

Key Takeaways

The Innovation Pipeline turns loose, half-formed ideas into focused, managed change. Instead of letting good concepts stall, it walks them step by step from spark to rollout. Each proposal is tested for real business value, fit with strategy, practical feasibility, and impact on the broader architecture before anyone spends serious time or money on it.

Enterprise Architecture is present from start to finish. Architects shape rough ideas into something concrete, surface hidden dependencies, connect efforts to roadmaps, and guard consistency across the portfolio. They see where pieces collide or overlap and fix them early. That steady involvement keeps the process from splintering and helps strategy actually turn into results.

The pipeline throws the doors open when gathering ideas, then tightens expectations as initiatives move forward. Oversight grows in proportion to risk and impact. Teams get room to think, but accountability never disappears. Strategy means nothing if it doesn't shape what gets built. The pipeline ties big ambitions to day-to-day delivery, making sure the products, systems, and capabilities the organization creates match the future it says it wants.

CHAPTER 30

Idea Generation

Ideas don't show up on a schedule. They surface in hallway chats, during project wrap-ups, inside budget reviews, or in that irritated moment when someone becomes convinced that there has to be a better way. Enterprise Architecture shouldn't try to manufacture brilliance on demand. It should build a simple way to catch it when it appears. An architecture team that waits for polished proposals will miss the messy, early sparks that actually drive change and those sparks are where the real value often hides.

In the Kettera EA Methodology, generating ideas means gathering raw "what if" thoughts before anyone cleans them up, prices them out, or writes a formal pitch. No slide decks or approvals. Just possibilities. At this point, an idea isn't a funded initiative or a promise to act. It's jus a question or idea. Could we automate this? Should that system even exist? What would happen if we did the opposite?

The discipline lies in capturing those questions early, before they're forgotten, dismissed, or reshaped into something safer. Ideas are fragile at birth. If you treat them like finished business cases, they'll never grow into recognized value.

Your architecture pipeline is only as good as the range and reliability of the raw inputs that feed it.

Ideas Can Come from Anywhere

One of the fastest ways to weaken an architecture function is to suggest that ideas must come from executives or architects. Innovation does not follow reporting lines. A customer service representative who handles complaints every day may spot inefficiencies that senior leadership cannot see. A database administrator might notice patterns that point to an entirely new data product. A finance analyst could identify overlapping reports that can be combined.

Ideas usually come from four places: day-to-day headaches, big growth plans, new tech showing up, and outside pressure from regulators or the market. If the top of your innovation pipeline is too narrow then imaginative ideas will be limited.

Friction on the ground sparks questions such as, "Why are we typing the same data into three different systems?" Big ambitions sound different like, "What if we switched to subscriptions instead of one-off sales?" New tools trigger their own push and lead to questions like, "Can we automate this?" And when rules change or competitors make a move, leaders ask, "How do we respond without hiring more people?" All are good questions driven by the constant change in our environments. Enterprise Architecture should handle these ideas as raw material, not polished solutions.

They're inputs to shape, test, and pressure-check.

The Mechanism Matters More Than the Spark

Most organizations aren't short on ideas. They're short on a reliable way to record and organize them. Conversations spark energy, whiteboards fill up, sticky notes multiply, and then it all fades into the rhythm of daily work. An EA cadence adds structure without stifling creativity. The idea system should stay simple, visible, and easy for anyone to use. That might mean a shared intake form, a central backlog, or simply a basic idea log. The specific tool matters far less than using it consistently.

You might set up a quarterly "architecture intake window" and invite teams to submit ideas tied to strategic priorities. Or keep submissions open year-round through the Innovation Pipeline mentioned in the previous chapter. Some organizations run focused ideation sessions right after strategy planning. Others place architects in regular operational meetings so they can spot recurring pain points as they surface. The common thread is this that ideas are written down in a standard format before they are reviewed.

Capturing Ideas in a Useful Way

Ideas on their own often lack clarity and specificity. They need a bit of shape before anyone else can use them. If you add too many rules at the start, people stop sharing but if you add none at all you end up sorting through chaos later. So be practical. An idea should come in with enough detail to make sense a week or a month later, but not so much paperwork that it feels like filing taxes. What problem does it address? Why does it matter now? What would success look like? Who might be affected and who might push back?

- A short title
- A brief description of the "what if"
- The problem or opportunity it addresses
- The expected benefit, even if it's only a rough estimate
- The department or individual submitting it

What's not included here? Detailed cost projections, rollout schedules, or technical blueprints. Those come later. At this point, the only question on the table is, "What if we tried this?", not "Exactly how will we carry it out?"

For example:

- **Title:** Consolidate Customer Data Platforms
- **What If:** What if we combined our three customer databases into a single source of truth?
- **Problem:** Sales, marketing, and support each keep separate records, creating inconsistencies.
- **Perceived Benefit:** Better customer experience and less time spent reconciling data.

That's enough to move ahead. It may or may not hold up after refinement, but it's now visible and easy to track. Capturing ideas this way helps architects spot patterns and take action in the future.

Encouraging a Culture of "What If"

Many companies don't mean to shut down creativity, but they do. The moment leaders expect a polished business case with spreadsheets included, people get the message. If it's not airtight, don't bring it up. After enough dismissals, employees stop sharing half-formed thoughts.

The EA cadence breaks that pattern by separating brainstorming from judgment. In this stage, nobody asks, "Can we build it?" The better question is, "Should we take a closer look?" That small shift in thinking changes everything. It gives people room to float possibilities without preparing a full defense. They can test a spark of an idea, even if it's rough around the edges, without fear of being shut down.

Leadership behavior makes or breaks this effort. When executives openly share their own unfinished concepts, they show that raw thinking is welcome. When architects reply with curiosity instead of criticism, more people join in. When leaders clearly explain why certain ideas won't move forward, trust will grow. When they thank contributors publicly, participation increases. Each action sends the same message: it's safe to think out loud.

One straightforward habit can make a big difference: regularly share a summary of the ideas that have been submitted. No scoring and no verdicts. Just a clear record of what came in. Employees see their input captured and discussed and they understand the process is real.

Linking Ideas to Strategy Without Restricting Them

Openness and alignment often pull in opposite directions. Ideas should be able to come from anywhere in the organization. At the same time, the business is working toward defined strategic goals. So how do you keep the backlog from turning into a grab bag of unrelated suggestions? The solution is to tag ideas rather than screen them out.

As ideas come in, link them to the strategic themes or capability domains described earlier such as marketing, operations, finance, IT, and similar areas. For example, an idea to automate invoice processing would fit within a finance capability. A proposal for a customer mobile app could connect to both marketing and application architecture.

Tagging ideas this way helps Enterprise Architecture see where energy is building up. It also reveals gaps. If no ideas connect to a stated strategic priority, that may point to a communication issue rather than a lack of opportunity.

Avoid rejecting ideas just because they don't neatly fit the current strategy. Some of the most valuable initiatives start as outliers. Strategy changes over time, and raw ideas often question existing assumptions. Sometimes, that's the point.

The Role of Architects During Idea Generation

Architects act more like guides than builders at this stage. They listen carefully, draw out details, and protect the conversation from jumping too fast into solutions. Clarity is the goal and not clever designs. When someone tosses out a fuzzy proposal like "Improve reporting," the architect doesn't nod and start sketching. They start asking. Which reports are we talking about? Who actually reads them? What choice is hard to make today because the right information isn't there?

Those questions refine the idea without boxing it in. The concept stays rough but no longer confusing. An architect holds back the instinct to open a diagramming tool or recommend a new system because structure comes later.

Early idea work is about shaping the problem, not fixing it. The architect's job is to make sure everyone understands what's really being asked before anyone talks about platforms, integrations, or data models. Why rush into blueprints when the foundation hasn't even been described?

Architects keep an eye out for overlap. When two departments float similar initiatives, the architect can combine them into one clear proposal or connect them before they drift apart. That prevents fragmentation from hardening into separately funded projects. They also stay neutral. This isn't the time to champion a favorite tool or swat down ideas that challenge current standards. Idea generation is about range, not restriction.

Practical Techniques for Structured Ideation

Spontaneous ideas matter, but structured sessions often lead to more useful ideas. Workshops tied to annual strategy reviews work especially well. Ask participants focused questions like:

- What processes take up too much time?
- Where do customers face delays or confusion?
- Which manual tasks could be automated in the next two years?
- What new services could we offer using capabilities we already have?

Prompts like these steer people's thinking without telling them what to conclude. Small groups made up of different functions usually spark wider-ranging ideas than meetings filled with one department. Put operations people in the room with IT and finance and hidden connections start to show up through dependencies no single team could spot on its own.

Pull out your current architecture documents like capability maps, application portfolios, data diagrams, integration flows and look at them as if they belonged to someone else. Ask a blunt question: If we were starting from zero today, what would we build instead? Why assume today's setup is untouchable? That challenge alone often unlocks daring suggestions because it gives people permission to stop tweaking around the edges and rethink the whole thing.

Make it a habit that after a major project goes live, set aside time to ask what new doors just opened. Now that this exists, what nearby opportunity can we grab? Finishing strong tends to reveal options that were invisible before because success has a way of changing the landscape.

When to Stop Generating and Start Refining

A healthy rhythm accepts that idea generation can't go on forever. At some point, the list has to settle so the real work of sharpening and examining ideas can start. Take a quarterly cycle as an example: the first month focuses on collecting ideas, the second on tightening and improving them, and the third on assessing value and setting priorities. When this schedule is set ahead of time, everyone knows when brainstorming ends and evaluation begins. Closing the intake window for a time also shows that the process is organized. Ideas aren't vanishing into a void. They're moving to the next stage. However, keep in mind that you can still have a mechanism capturing spontaneous ideas but the ideas may be added to a future review cycle.

This natural handoff prepares readers for the next chapter. Raw ideas that are captured, clarified, and tagged, are necessary but not enough. They're still rough. Some will overlap; others will be vague. A few will be unrealistic. Before value can be estimated or resources given, the ideas need to be refined.

Key Takeaways

Idea generation inside the Enterprise Architecture cadence runs on open doors and steady guardrails. Anyone, anywhere in the business can spark something worth exploring. The capture process should be dead simple, used the same way every time, and easy for everyone to see. Early inputs aren't meant to be slick decks or business cases. They're rough "what if" thoughts that might be half-formed.

Keep the intake process light. A basic frame helps people jot ideas down in a way that's useful later without scaring them off. Labeling each idea to a strategic theme keeps it tied to what matters while leaving room for imagination. Architects don't judge at this point, they guide. They sort, clarify, and connect dots, but they hold back from critiquing too soon.

Done right, this step turns random hallway chatter and scattered notes into one shared list of real possibilities. That running list feeds the next stage which is Idea Refinement, where the strongest concepts get tested, refined, and shaped into practical actions the business can actually take.

CHAPTER 31

Idea Refinement

When an idea arrives at the Refinement stage, it isn't starting from scratch. It already has energy behind it. The previous chapter explained how concepts get into the system in the first place: workshops, strategy meetings, day-to-day operational frustrations, and new technology suggestions. Those inputs add up quickly and, before long, there's a robust backlog. Some ideas are ambitious. Some are half-formed. A few directly clash with others. Every single one is asking for time, money, and focus. Now we have to determine which of them deserves to move forward?

This stage exists because raw creativity isn't the problem, discipline is. A long list of possibilities might feel productive, but without organizing and sharpening them, it's just noise. The organization now has to look closely, and separate what sounds good from what can actually work.

Idea Refinement

Architecture Backlog Refinement turns rough, half-formed ideas into candidates that can actually be evaluated. Not approved projects. Not funded initiatives. Just feasible ideas, clear, scoped, and grounded enough to be assessed in the next stage. When teams skip this step, value analysis turns into guesswork but when they take the time to refine them, decisions rest on clarity instead of excitement.

From Raw Ideas to Architectural Candidates

Most ideas enter the architecture backlog half-formed. A business executive might say, "We need better customer analytics." A technology leader might suggest, "Let's modernize our integration platform." An operations manager might say, "We need to automate this workflow." Each statement points to something real, but none are ready to act on.

Refinement starts with a few direct questions: What problem are we solving? Who is affected? What outcome do we expect? What constraints apply? These questions are not bureaucratic hoops; they act as filters that prevent confusion later. If an idea cannot survive basic clarification, it should not move forward. Clarity comes first.

Consider the "better customer analytics" example. Does that mean improved dashboards? A new data warehouse? Predictive modeling? A customer 360 platform? Or is the real issue inconsistent data definitions across departments? Refinement helps separate symptoms from root causes. Sometimes the idea shifts completely once the true issue becomes clear but remember that the goal is not to reject ideas but to improve them.

The Architecture Backlog as a Managed Asset

In the Kettera methodology, the architecture backlog is treated as a managed asset and not a dumping ground for half-formed ideas. Each item must meet a clear minimum level of definition before it can move forward.

At regular refinement sessions, usually held on a steady cadence, the Enterprise Architecture team partners with idea sponsors to shape proposals into a standard structure. The goal is consistency so every idea can be reviewed on equal footing. A well-formed entry typically contains:

- A concise problem statement
- Desired business outcome
- Scope boundaries
- Dependencies and constraints
- Preliminary impact areas (business, application, data, integration, platform)
- Risks or assumptions

What you will not see at this stage is a detailed solution design. Refinement does not mean designing the answer. It means framing the idea clearly enough that it can be assessed across the enterprise. This structure creates comparability. When every idea is framed the same way, leadership can evaluate them fairly. A proposal doesn't move forward just because it was delivered more persuasively.

Separating Opportunity from Impulse

A rival rolls out a flashy feature or a software vendor promises amazing things in a demo. A senior leader skims a piece about artificial intelligence and wants something done by next quarter. Reactions like these are human but reacting isn't the same as thinking. Refinement means slowing down and asking harder questions.

The Enterprise Architecture team acts as a kind of ballast, the weight that keeps the ship from tipping over. They press on the basics: Does this idea truly support where the organization is headed? Do we already own something that solves this? How does it fit with projects underway? What will we have to stop doing to make room for it? At times, the wisest answer sounds dull: "We've got this covered already, just in a different way."

Picture a department asking to buy a new workflow automation tool. It looks sleek and the pitch is convincing. During a closer review, the architects uncover something: the company recently paid for a broad integration platform that already includes workflow features, and many of those features are barely touched. So what's the real issue, a missing tool, or a neglected one? The better move may be training people and using what's on hand rather than signing another contract. This isn't red tape. It's making sure the pieces fit together and move in the same direction.

Defining Feasibility Early

Here, feasibility does not imply full technical validation as that level of scrutiny comes later. At this stage, refinement is about filtering out ideas that are plainly unrealistic or out of step with the organization's direction.

Early feasibility usually hinges on three factors:

1. **Strategic Fit** – Does the idea advance the organization's stated goals?
2. **Architectural Alignment** – Is it consistent with existing standards and reference frameworks?
3. **Operational Practicality** – Can the organization realistically carry it out given expected constraints?

Take an organization committed to a cloud-first strategy with a designated vendor. If someone proposes building a new on-premises data center, questions would surface immediately. Without a strong strategic justification, the proposal would likely be rejected for misalignment before any deeper technical review begins.

Early feasibility checks prevent wasted effort and save time. They also build trust in the process. Stakeholders know their ideas will be considered seriously but not accepted blindly.

Right-Sizing the Scope

Some proposals try to cover far too much ground. "Digitally transform customer service" sounds impressive, but it's a direction, not a real project. You can't assign it, budget it, or finish it. Breaking that big ambition into smaller, workable pieces makes it real. Maybe the first step is rolling out a single case management system across teams. Maybe it's pulling all customer contact data into one place so everyone sees the same information. Those are concrete initiatives people can actually deliver.

Other requests go the opposite way. They zoom in so tightly that they miss the bigger problem. Automating one report might look harmless, but why does that report need attention in the first place? The pain could point to messy data rules or a tangled reporting setup which are deeper issues hiding under a small fix. When that happens, refinement widens the view and aims at the foundation instead of patching the crack. That new scope matters. If it's too big then nothing starts but if it's too small, then nothing really changes. The goal is to land in the middle where the work is focused, meaningful, and actually doable. Architects have to use sound judgment here. Overengineering is a constant temptation, and under-scoping is just as common. Knowing where to draw the line comes with experience.

Clarifying Impact Across Domains

One reason Enterprise Architecture exists is to look across boundaries. Every idea, even one that seems isolated, has ripple effects. A new mobile application affects data management, integration patterns, security standards, and operational support. A finance reporting enhancement can change data definitions used by marketing and operations. During refinement, architects outline these impacts at a high level.

Looking across domains does two big things. It keeps surprises from popping up late in the game, and it uncovers chances to combine efforts. Separate proposals often depend on the same data integration fix, sometimes without anyone noticing. Why implement them separately? When leaders spot that shared need early, they can tackle it once, in a coordinated way, instead of burning time and money on parallel tracks.

Refining ideas isn't just about tidying up each proposal on its own. The real work is understanding how those ideas connect, overlap, and sometimes collide across the organization, how one shift in data, systems, or priorities ripples outward. That broader view turns a stack of good suggestions into a coherent plan.

Engaging the Right Stakeholders

Architecture Backlog Refinement doesn't happen in a vacuum. The Enterprise Architecture team runs the sessions, but idea sponsors are still responsible for clearly stating the business value and intent behind their proposals. Strong refinement sessions bring in voices from across affected areas like business leaders, application owners, data architects, security teams, and at times finance. The aim isn't to walk out with final approval; it's to raise concerns early, while there's still room to adjust.

A security architect might point out regulatory implications. A data architect might question data quality assumptions. An operations leader might raise support capacity constraints. Catching these issues in refinement prevents bigger problems later. There is also a cultural benefit. When stakeholders see that their perspectives are taken into account at this stage, resistance drops in later phases.

Documenting Without Overburdening

During refinement, documentation needs structure and focus, but it shouldn't balloon into a paperwork exercise. In most cases, two well-organized pages will do the job. The goal is simple: make the thinking clear, not create a process people resent. Too much detail at this point drains time and energy but too little leaves too much ambiguity. The sweet spot is capturing just enough information to compare options, weigh tradeoffs, and move forward with confidence without sliding into full-blown design mode.

Architecture Decision Records, introduced earlier in the book, can also be cited when an idea pushes against current standards. If it departs from a reference framework, say so. Refinement won't fix that conflict, but it makes the tradeoff visible before more time or money is spent.

Clear, consistent documentation builds institutional memory. Over time, the organization can look back at refined ideas and see the reasoning behind each decision.

Knowing When an Idea Is "Refined Enough"

An idea is refined enough when:

- The problem and desired outcome are clearly spelled out.
- The scope limits are defined.
- Key dependencies and constraints are visible.
- Strategic fit has been reviewed.

• The idea can be fairly compared to others in the backlog.

Keep polishing forever and you'll stall out in analysis paralysis. Stop too soon and the next steps become more difficult. The rhythm laid out in Chapter 27 keeps things moving without spinning wheels. Set up steady refinement sessions and the backlog will stay alive and ideas get better bit by bit but how long can a team afford to think before making a decision?

A Practical Example

Take a mid-sized financial services firm that puts forward a simple idea: "Implement AI to improve underwriting."

As the team examines it more closely, a clearer picture forms. Risk assessments differ from one regional team to another, which slows approvals. Data quality is uneven. There's also no centralized underwriting rules engine in place. An "AI solution" would depend on standardized data inputs, yet those inputs don't exist today.

After working through these gaps, the idea shifts to: "Establish a centralized underwriting rules platform and standardized risk data model, enabling future advanced analytics." The focus tightens, and key dependencies come into view. Instead of chasing buzzwords, the organization now has a realistic initiative it can actually execute.

Key Takeaways

Idea Refinement turns enthusiasm into discipline. Through Architecture Backlog Refinement, the organization turns loosely defined suggestions into clear, workable candidates. This process defines the problem, right-sizes the scope, checks fit, reveals dependencies, and records ideas in a consistent way.

By the time this polishing work wraps up, every idea can stand on its own and be judged on its merits. Each one is clear enough to judge without assumptions. Some will move forward and some will not but that is how the process is designed. The purpose of Enterprise Architecture is not to get every idea implemented but to help determine which ideas are implemented.

CHAPTER 32

Value Analysis

An idea that reaches value analysis has already cleared the first hurdles. It more or less fits the strategy. People understand it. The problem is clear, and the proposed fix makes sense. Now comes the harder question: is it actually worth the time and money?

Value analysis turns a backlog of promising concepts into a short list of ideas that make financial sense. In the Kettera Enterprise Architecture rhythm, this step serves as financial component of backlog refinement for architecture work. Concepts move out of the "sounds good" stage and into real scrutiny: costs, impact, effort, risk. The goal isn't to label something a good idea. The goal is to prove it delivers measurable value, with a return you can explain and defend.

From Idea to Investment Candidate

Value analysis starts by clarifying the intended outcome in

measurable terms. If an initiative claims to "improve operational efficiency," how will that be measured? Shorter cycle time? Fewer manual steps? Lower labor cost per transaction? A 15% decrease in ticket resolution time is measurable. "Better support experience" is not.

This is where architecture adds financial sense. Each idea is translated into expected financial impact, whether direct or indirect. Revenue increases, cost reductions, risk avoidance, and productivity gains are all forms of economic return. Even initiatives driven by compliance or modernization must ultimately be framed in financial terms. What happens if we do not act? What cost will continue, grow, or eventually be imposed?

The discipline lies in turning abstract benefits into economic terms. It requires collaboration with finance, operations, and technology teams. Architecture does not guess at value. It models it.

Return on Investment as the Primary Lens

ROI is the main yardstick. Other factors such as strategic fit and risk count but ROI gives everyone a shared way to judge value. It turns a messy mix of proposals into numbers you can line up side by side and compare with objectivity. At its core, ROI measures what you get back versus what you put in. Spend $500,000 and generate $1.5 million in net gains over three years. That's a strong return. It's simple math on the surface but there is real complexity underneath.

Numbers alone don't tell the story. When do the benefits

actually show up? Do the savings repeat year after year, or do they happen once and disappear? Which assumptions are propping up the revenue forecast and how fragile are those assumptions? Small shifts in those inputs can swing the outcome in meaningful ways.

Enterprise architecture should demand transparency. When projected savings depend on a 30% productivity gain, where did that number come from? Was it drawn from industry benchmarks, pilot results, or simple optimism? A disciplined value analysis brings those assumptions into plain view.

Take a plan to merge three legacy systems into one modern platform. The upfront costs are significant. Licenses, implementation, migration, and training add up quickly. That price alone can make stakeholders pause. But a broader picture tells a different story: ending old maintenance contracts, simplifying integrations, removing overlapping support roles, and improving data accuracy all contribute to long-term returns. Without a structured review of value, those benefits stay hidden behind short-term sticker shock.

ROI reframes the discussion. Instead of asking, "How much will this cost?" the better question becomes, "What return do we get for what we put in?"

Understanding the Full Cost of Change

Sound ROI calculations rely on accurate cost modeling. Underestimating costs is one of the most common mistakes in initiative planning. Architecture must account for more than software licenses or hardware purchases. Total investment includes implementation effort, integration work, data migration, process redesign, training, change management, and ongoing operating expenses. It also includes opportunity cost, the value of other initiatives that will not be pursued if this one is approved.

For example, automating a manual financial reconciliation process may involve modest software licensing fees but significant integration work with existing systems. If integration complexity is ignored, projected ROI will be inflated. Enterprise architects are well positioned to spot these hidden costs because they see cross-system dependencies that individual teams may miss.

At the same time, value analysis should not pad costs out of caution. Overly conservative estimates can shut down worthwhile initiatives. The goal is balance; realistic projections grounded in architectural understanding.

Quantifying the Benefits

Benefits show up in different ways, and you have to spell out each one in plain terms. Don't assume people will connect the dots for you. What exactly gets better, and how do you know? Lay it out so anyone can see the gain.

Cutting costs is the easiest place to start. When you shut down duplicate systems, trim the number of vendor contracts, use less infrastructure, and replace manual tasks with automation, the savings are real and trackable. Add them up for a single year, then extend the math across a set number of years so leaders can see the full impact. Spell out the time frame, show your assumptions, and make the totals easy to follow.

Revenue growth is often harder to pin down. A new digital capability can speed up product launches or help the company reach entirely new customer segments, but putting numbers to that potential takes work. Business leaders need to weigh in, and teams usually build scenario models to test different outcomes. Looking at conservative, moderate, and optimistic projections helps clarify both the upside and the level of risk involved.

Risk reduction also deserves close attention. Investments in security, compliance, and system resilience may not bring in new revenue, yet they shield the organization from serious losses. Their value shows up in costs avoided. For example, if a system outage has historically cost $200,000 per hour and an architecture improvement cuts the likelihood of outages in half, the expected financial gain can be calculated. Once risk is expressed in financial terms, it moves from being a vague concern to a measurable part of ROI.

Some benefits, such as improved customer satisfaction, are harder to measure. Yet even these can often tie back to clear results like retention rates, upsell performance, or lower support costs. Architecture connects intangible claims to real metrics.

Architecture Backlog Refinement in Practice

Teams run focused review sessions to arrive at a meaningful value analysis. Each backlog item comes with a short value brief that lays out the problem, the proposed fix, expected costs, and the benefits the business should see. That's enough to ground the conversation without drowning in paperwork.

During these sessions, architects engage with focused questions: Which dependencies are adding complexity? What capabilities already exist that could lower costs? Can the work be staged to deliver value earlier? Would a small pilot test key assumptions before committing full funding?

At times, this analysis shows that a sweeping initiative should be divided into smaller steps. Rather than replacing an entire platform in one effort, an organization might upgrade the most expensive components first and use the early savings to support later phases. When value arrives sooner and risk drops, ROI tends to improve as well.

Avoiding the Trap of Vanity Projects

Every organization collects great ideas that promise excitement but little real economic return. A new analytics dashboard that few people will use. A technology upgrade driven by trend rather than need or a pilot that never scales. Value analysis shields the organization from these distractions. If the projected ROI is minimal or highly speculative, the initiative stays in the backlog or is cut entirely. This discipline can feel uncomfortable, especially when senior leaders are attached to an idea but governance requires consistency.

Not every project has to deliver a splashy payday. Some efforts lay the groundwork for what comes later, setting up gains that show up down the road. That's fine. But even those building-block investments need clear, stated outcomes. If standardizing data allows three divisions to launch products faster, then estimate the total lift in revenue or savings that speed creates. Foundational doesn't equal impossible to measure. It just means the payoff may take longer to see.

Anchoring conversations in ROI keeps big personalities and office politics from steering the ship. When numbers are on the table debates shift away from opinions and toward evidence. Clear financial framing turns architecture decisions into clear business decisions and reduces the pull of personalities.

Time Horizon and Value Realization

Not all value arrives on the same schedule. Some efforts cut costs quickly but deliver little staying power, while others take time before the payoff becomes visible.

When assessing ROI, enterprise architecture teams need to look across the right time frame, typically three to five years. A narrow, short-term lens can punish initiatives that are meant to reshape the business. If we stretch projections too far out, though, the numbers will become speculative.

Payback period adds another perspective: how long before total benefits surpass the upfront investment? Faster payback lowers financial exposure and tends to build leadership confidence.

For example, implementing robotic process automation in accounts payable can deliver measurable savings within a year. Replacing a core platform could take two years before it shows net benefits. Both may be justified, but understanding the timing helps shape a balanced portfolio.

A healthy backlog includes a mix of short-term returns and long-term strategic investments. Value analysis makes that balance clear.

Producing Valuable Ideas

After value analysis, the architecture backlog is transformed. Vague intentions become clear. Every remaining idea now shows its estimated cost, expected benefits, ROI, payback timeline, and the assumptions behind the numbers. Instead of loose concepts, you have proposals grounded in math and stated estimates.

This cleaned-up backlog sets up the next stage: prioritization. With ROI and value spelled out, comparisons get easier and far less emotional. Teams can debate real trade-offs with full transparency. Limited people, time, and money flow toward the opportunities with the strongest returns. The conversation shifts from opinion to evidence.

Keep in mind that predictions aren't perfect. Assumptions can be too hopeful or too cautious. Markets will shift. Technology keeps moving, sometimes faster than anyone expects. Even so, putting numbers around value forces rigor and clarity. It strengthens the role of enterprise architecture as a guardian of investment decisions, not just a group that designs systems.

Enterprise architecture keeps its seat at the leadership table by talking about value. Diagrams show how things fit together, roadmaps point to where the organization is headed, and ROI explains why a choice makes financial sense. When leaders hear that language, the work stops sounding abstract and starts sounding practical.

The next chapter looks at how these ideas are weighed against each other. Once you understand the value of each option, the next question is which initiatives should come first?

CHAPTER 33

Prioritization

A backlog filled with promising ideas might feel productive, but it's only the beginning. Nothing changes until we act on those ideas. Enterprise Architecture matters when leaders decide what gets addressed now, what can wait, what deserves more study, and what stops here. This is where strategy collides with budgets, politics, timing, and actual capacity. Prioritization is that collision point.

In the Kettera EA Methodology, prioritization comes after ideas are generated, shaped, and tested for value. That order is deliberate. By this point, rough thoughts have been solidified, weak assumptions have been tested, and the expected benefits are visible. Still, analysis doesn't make anything happen. What turns thinking into motion is a clear system that names the most important Ideas right now. Most organizations rely on a cross-functional steering group to do this work, meeting on a regular cadence and working from shared criteria, so decisions are based on a sound and consistent foundation.

From Value to Decision

Value analysis asks, "Is this idea worth doing at all?" Prioritization asks something tougher: "Is it worth doing more than the other options right now?"

One initiative might promise a high return, fit neatly with the current architecture, and improve day-to-day operations. Another could lower risk in a tightly regulated area. A third might open up a new revenue stream but demand major change management. Each makes sense on its own. The strain shows up when money, people, and time are scarce, as they always are.

Effective prioritization requires comparative thinking. Ideas must be weighed against each other, not evaluated in isolation. That comparison includes quantitative measures like expected financial return or cost savings, as well as qualitative factors like strategic alignment, architectural coherence, risk reduction, customer impact, and organizational readiness. The goal is not mathematical precision. It's disciplined judgment.

Enterprise Architecture plays a unique role here. Architects see the cross-domain effects like how a marketing initiative may require changes to the data model, how a finance automation effort affects integration patterns, how a platform upgrade shapes application lifecycles. Without that broader view, prioritization often slips into departmental lobbying. With it, the discussion shifts from "my project" to "our enterprise trajectory."

The Role of the Cross-Functional Steering

Committee

Handing enterprise-wide priorities to a single executive invites tunnel vision. Splitting decisions across disconnected teams creates chaos. A cross-functional steering committee sits in the middle and keeps both extremes in check.

Membership usually spans senior leaders from marketing, operations, finance, and HR, plus IT leadership and Enterprise Architecture. The blend is very important. Business heads understand strategy and what customers actually care about. IT leaders know what can be built, how fast, and at what strain on the system. Enterprise architects look across the whole landscape and guard long-term fit. Product leaders often add a dose of market reality.

The job is simple: review vetted initiatives and rank them for the next planning window. That window might follow quarterly funding, annual budgets, semiannual reviews, or a rolling 12- to 24-month roadmap. The exact timeframe matters less than the rhythm. Without a steady cadence, priorities swing toward whoever shouts loudest or whatever recent crisis is grabbing people's attention.

Picture five proposals that have already cleared value review: upgraded customer analytics, modernization of a core platform, a regulatory reporting overhaul, a new digital channel, and automation for internal workflows. All are worthwhile. In a structured session, the committee lines them up against agreed upon criteria, debates the trade-offs in the open, and orders them. The result isn't just a ranked list. It's shared clarity about why some initiatives move now and others wait, and everyone walks out understanding why.

Criteria and Transparency

Teams typically weigh factors such as strategic alignment, financial return, risk reduction, architectural fit, customer impact, and implementation complexity. Those weights should not stay fixed. In a regulatory crunch, reducing risk may take priority; in a period of expansion, revenue growth may matter more. The steering committee needs to agree on these trade-offs at the start of each planning cycle instead of shifting them later to defend preferred projects.

Clarity should not stop with the committee. People across the organization need to see how and why decisions happen. When an initiative is postponed, stakeholders deserve a specific reason, not a vague statement about "timing." A short explanation such as, "This initiative aligns with our strategy but will follow the core system upgrade because of integration dependencies," preserves credibility.

Enterprise Architecture plays a key role by exposing dependencies. A roadmap that shows sequencing constraints, technical debt, and capability maturity makes the limits visible. Without these artifacts, choices can seem arbitrary; with them, trade-offs are tangible and easier to discuss.

Managing Politics and Bias

No prioritization process is free from politics. Leaders advocate for their domains, and incentives shape behavior. Pretending otherwise is naïve. The goal is not to eliminate politics entirely but to channel it in a constructive way.

Structured scoring models can help by offering a shared framework. However, rigid formulas can create a false sense of objectivity. A steering committee should treat scoring as input, not a final verdict. Discussion still matters. Why does one initiative score lower on complexity? Are assumptions consistent across proposals? What risks have been understated?

An experienced chairperson, often a senior executive or the CIO, plays a vital role in guiding the discussion. Strong personalities cannot be allowed to dominate. Quieter but strategically important voices must be heard. Enterprise architects can act as neutral advisors in these conversations, pointing out cross-functional impacts without pushing departmental agendas.

At times, two initiatives will appear nearly equal in merit. In those cases, sequencing, resource availability, or organizational fatigue may tip the balance. A company already in the middle of major change may decide to stagger transformation efforts rather than launch several disruptive programs at once. Prioritization is as much about timing as importance.

Output: The Most Important Ideas

A prioritization exercise does not produce just another spreadsheet. It produces organizational focus. What the organization truly needs is a clear, ordered list of initiatives, the few that matter most in the next stretch of time.

That list usually separates work into four buckets: "Approved for Execution," "Deferred but Retained," "Under Active Review," and "Not Pursued." Approved items move straight into staffing plans, budgets, and formal project sign-off. Deferred ideas stay on the radar and come back up in the next planning cycle. Items under review need more facts before a call gets made. The ones not pursued are stored with a written explanation, so they don't keep popping up every few months without any new reason to reconsider.

This clarity changes behavior across the organization. Teams know what deserves their energy. Budget talks rely less on opinion and more on shared priorities. Outside partners hear the same message from everyone. Architecture benefits most of all. When decisions reflect the needs of the whole enterprise, not just one team's wish list, the collection of projects pulls the organization toward its intended future instead of creating scattered wins that don't add up.

Consider a practical scenario. The steering committee decides to prioritize modernizing the core system instead of rolling out a new digital channel. That choice may frustrate the marketing team, which is eager to boost the company's external profile. But when leaders explain that upgrading the core will allow faster product configuration, cleaner data, and reduced operational risk, the rationale becomes clearer. In this light, the decision reflects a longer-term architectural path rather than a dismissal of innovation, even if it doesn't feel that way at first. The digital channel isn't abandoned; it's simply scheduled to follow once the necessary capabilities are in place.

Maintaining Cadence and Discipline

Within the Enterprise Architecture cadence, prioritization sessions take place at regular intervals, often quarterly. Between sessions, new ideas enter the pipeline and existing initiatives may shift in scope or urgency. The steering committee reviews these changes within the established structure, avoiding reactive decisions driven only by outside pressure.

Once a decision is made, treat it as real. Shuffling priorities every few weeks chips away at trust and throws teams off balance. People stop believing the roadmap means anything. Work slows down and focus fades. Urgent issues will frequently pop up such as new regulations or a serious competitive move, but recognize them as exceptions. Identify them clearly and handle them carefully. If everything becomes urgent, what was the point of prioritizing in the first place?

Enterprise Architecture provides steady oversight from one planning cycle to the next. Architects keep track of how each approved initiative moves the organization toward its target state. When someone suggests changing direction, they can show the ripple effects such as higher long-term costs, disconnected systems, a patchwork landscape that's harder to manage later. Their role is simple but critical: prevent quick wins from steering the company off course. Without that steady view, short-term fixes add up, and the bigger vision slowly falls apart.

Cultural Impact

A disciplined approach to prioritization doesn't just organize projects, it gradually reshapes culture. Leaders start weighing enterprise-wide value instead of protecting departmental turf. Sponsors frame proposals with cross-functional effects in mind, knowing that narrow cases rarely succeed.

Over time, people gain confidence that strong ideas will receive fair consideration. Even proposals that are postponed benefit from a clear process; the act of review itself builds trust. It's far easier to accept "not now" than to hear "no" without any context. That trust strengthens the broader Enterprise Architecture effort. EA is no longer viewed as an abstract practice that produces diagrams; it becomes part of everyday decision-making. Architects are not bystanders, they help shape what the enterprise does next.

Key Takeaways

Prioritization turns a collection of analyzed ideas into a clear order of action. It's about weighing options against each other, not judging each one on its own. A cross-functional steering committee is the right vehicle for this work, bringing together people who can weigh big-picture goals, day-to-day reality, long-term technical fit, and customer impact.

Clear standards matter but open communication matters more. When people understand how decisions are made, we inspire trust. A scoring model can guide the conversation, but it should never silence it. The real result is a ranked list of the Most Important Ideas, ready to move into the next phase of the cycle and rally the company around a shared focus.

Now we have to determine if we actually have the time, budget, and people to pull this off? Once priorities are set, that question can't be ignored. It drives the conversation into resource planning, where strategy connects with hiring plans, financial limits, and the calendar.

CHAPTER 34

Resource Planning

Resource planning estimates the people, skills, funding, and timing required and checks those needs against the organization's boundaries: Capacity, Capability, Cost, and the trade-offs between them.

From Priority to Practicality

Prioritization answers the question, "What should we do next?" Resource planning answers, "What can we realistically do next, and how?" These are not the same. A high-priority initiative might require skills the organization doesn't have, depend on vendor contracts that aren't yet in place, or need funding beyond current budgets. Ignore these realities and the roadmap remains nothing more than a good idea that can't be executed. Teams will commit to timelines that fall apart under hidden constraints, and trust in the architecture function erodes.

Resource planning is essentially a structured form of backlog refinement. It looks at each prioritized initiative and checks it against available people, technical skills, budget limits, and vendor commitments. The goal isn't to shut down ambitious ideas, but to understand the conditions needed to deliver them.

You get two clear deliverables from this work. One is a practical roadmap that matches priorities to real-world capacity across the next 12 to 24 months that the team can actually handle. The other is an implementation vision that spells out exactly what's needed, laid out through the lens of the Three C's; capabilities, capacity, and capital.

The Three C's in Practice

Earlier in the book, we described the Three C's—Capability, Capacity and Capital, as architectural constraints. Resource planning is where those ideas become more visible.

Capability looks at whether the right skills are in place. Do we have seasoned cloud architects? Data engineers who truly understand the current warehouse? Business analysts who can translate regulatory requirements into system design? Gaps often become visible here, especially when new technology or operating models enter the picture.

Capacity is about how much work the organization can realistically take on. It includes staff bandwidth, other active initiatives, ongoing operational demands, and the very real issue of change fatigue across teams. Even a well-designed solution will stall if people are already overloaded and have no room to implement or adopt something new.

Capital extends well beyond a simple budget line. It covers direct spending such as licenses, hardware, consulting support, as well as indirect impacts like training time, transition slowdowns, and the temporary dip in productivity that tends to follow major change.

When resource planning is disciplined, every initiative is evaluated clearly against these three dimensions. If an initiative scores high in value but goes beyond current capacity, the roadmap must adjust. If capability is lacking, leadership must decide whether to build internally, hire, or bring in external partners. If available capital is too low, either the scope must change or funding strategies must be rethought.

Building the Implementation Vision

An implementation vision is not a Gantt chart with tidy bars and dates. It tells a story. It shows, clearly and visually, how the most important initiatives will actually get done, given the real limits of people, money, and time inside the organization.

Picture a company that has decided to modernize its customer relationship management (CRM) system and roll out advanced analytics to improve sales forecasts. Both promise meaningful returns. Looking at them one by one is easy but looking at them together is harder. The same developers may be needed for both. The data team can't be in two places at once. Budgets stretch only so far. So planning shifts from isolated projects to a shared, realistic view of what can move forward and when. This is the heart of the vision. It forces leaders to ask a simple question: given our constraints, what can we truly deliver, and in what order?

Modernizing the CRM will likely demand substantial integration work along with thoughtful change management. The analytics effort relies on data engineering expertise that is currently in short supply. Consolidating finance systems could require bringing in outside vendors and renegotiating contracts. Taken together, these efforts draw on many of the same technical specialists and business leaders, creating real competition for limited resources.

A clear implementation vision helps determine what happens first. The organization might launch the CRM upgrade right away, hold off on analytics until core data capabilities are stronger, and time the finance consolidation to coincide with the start of a fiscal year to ease budgeting pressures. The document should spell out not just the order of initiatives but the reasoning behind it, showing how the plan accounts for available capacity, existing skills, and budget limits.

This level of clarity avoids confusion down the line. Stakeholders understand why something is planned for Q4 instead of Q2. The roadmap is no longer just a list of dates; it becomes a story about how the value will be delivered.

Internal Constraints: The Reality of Capacity

Many companies get the question of capacity wrong. The org chart says people are free. The project plan says capacity exists. Day to day, though, those same teams are buried in support tickets, compliance work, and promises they already made months ago.

Enterprise architects need to dig into what capacity is really available. How many big efforts can the sales team actually handle in six months before performance slips? Are your go-to experts already tied up on other initiatives? How long does it truly take for a new hire to get productive? And who is tracking all of this in one place?

Capacity isn't purely a technical matter. Business units have limits too, and they're often the first to feel the strain. A transformation that calls for major process redesign can easily clash with peak seasonal demands. A retailer, for instance, should never roll out a new order management system on the eve of the holiday rush, even if it aligns perfectly with the long-term architecture.

Resource planning works best when these constraints are acknowledged upfront. Sometimes that means cutting back the number of active initiatives to maintain focus. Doing so can be uncomfortable, especially when leadership enthusiasm is running high, but it prevents fragmentation and keeps teams from burning out.

Capability Gaps and Strategic Choices

Capability constraints often surface deeper strategic questions. If a top priority requires advanced machine learning expertise the organization doesn't have, leaders face some choices: develop internal talent, hire specialists, partner with a vendor, or postpone the initiative. Each choice has consequences. Building internal capability takes time but builds long-term independence. Hiring may speed up delivery but increases fixed costs. Engaging a vendor can provide immediate expertise, yet creates reliance on external partners and potentially higher long-term costs.

Resource planning does not dictate the answer. It makes the trade-offs explicit. The implementation vision might state: "This initiative requires cloud security expertise currently unavailable internally. We will engage a specialized vendor for initial deployment while training internal staff to assume operational ownership within 12 months." That statement aligns capability planning with architectural intent. It links today's delivery needs to tomorrow's operating model.

External Constraints and Vendor Dependencies

Almost no major project lives entirely inside your walls. Outside players like software companies, integration firms and cloud providers, shape how fast you can move and how much you will spend. Their product roadmaps, pricing structures, contract fine print, and support policies can either speed things up or box you in.

Picture a company preparing to roll out a new ERP module. The vendor might restrict deployment to certain release windows. Pricing terms could push the business to buy several modules together, even if it only needs one right now. Implementation partners may already be booked solid for parts of the year. How much control do you really have if those pieces don't line up?

Planning without these constraints sets the stage for frustration. Plans built in a vacuum rarely survive first contact with vendor schedules. A grounded implementation vision spells out the outside forces at play such as vendor delivery dates, required third-party support, contract restrictions, and the risks that come with relying on someone else's performance. Clarity here prevents wishful project blockers later.

Architects also need to think about concentration risk. When several initiatives hinge on a single vendor, a delay on their side can stall everything. That's why the roadmap should spell out backup options or a plan to spread dependencies across multiple partners.

Resource planning isn't just an internal headcount exercise. It has to account for the wider network the enterprise relies on such as suppliers, contractors and service providers.

Integrating Roadmaps and Backlog Refinement

At this stage in the cadence, the backlog still isn't fixed. Resource planning can lead to reordering or adjusting scope. An initiative that seemed straightforward may uncover hidden integration complexity. Another might turn out to be easier than expected because of reusable components or shared capabilities. This kind of iteration is healthy. It shows that the architecture function is actively involved, not just documenting decisions made elsewhere.

The refined roadmap should clearly show sequencing across domains, including business, application, data, integration and technology. Dependencies need to be clearly visible. For example, a data governance initiative may need to come before expanding analytics. A platform upgrade might be required before launching new customer-facing features.

When leaders can see this clearly, they understand how corporate or departmental strategy connects to the work getting done on the ground. The link between plans and execution becomes visible and concrete. Better visibility also leads to smarter budgeting and more realistic staffing plans. Instead of guessing, executives can match resources to actual priorities.

A Practical Example

Consider a mid-sized financial services firm mapping out a two-year roadmap. Through prioritization, leaders settle on five major initiatives: digital onboarding, enhanced fraud detection, a core banking upgrade, a data governance program, and infrastructure modernization.

As planning gets underway, competition surfaces. The same core banking team is needed both for the upgrade and for integrating digital onboarding. The data governance effort calls for experienced data stewards, yet those people are tied up with regulatory reporting. At the same time, infrastructure modernization collides with upcoming vendor contract renewals.

So instead of launching all five at once, the company sequences the work with care. Year one centers on the core banking upgrade and essential infrastructure updates, backed by selective vendor partnerships. Digital onboarding moves forward too, but with a narrower scope. Improvements to fraud detection are scheduled for year two, after stronger data governance is in place.

Some stakeholders are likely to feel frustrated at first. Even so, the plan matches ambition to available people and skills, and it presents a clear path of change rather than a patchwork of unrelated projects. Whatever frustration is experienced up-front due to the sequencing will be absorbed better than the frustration experienced during the implementation due to mid-flight delays.

The Discipline of Saying "Not Yet"

Resource planning takes restraint. Some projects must wait and others need to be redesigned to fit within current limits. Saying "not yet" is different from saying "no." When the reasoning is clear and grounded in the Three C's, stakeholders are more likely to accept the delay. They can see that waiting protects delivery quality and organizational stability.

Over time, this discipline strengthens the credibility of the architecture function. Roadmaps are delivered as promised. Resource conflicts are foreseen rather than discovered mid-project and trust will grow.

Preparing for Formal Approval

Resource planning should leave the organization with a well-defined, usable picture of how work will roll out. A practical roadmap backs it up. Every initiative is sequenced in a clear order, with its needs spelled out in terms of capacity, skills, budget, and timing. Limits, both inside the company and outside it, are spelled out and trade-offs are clear.

This is what makes formal approval possible. Executives shouldn't be hearing about staffing gaps or budget strain for the first time in the approval meeting. Why would they sign off on something they're seeing for the first time? Approval becomes a simple confirmation of a plan already refined and understood.

CHAPTER 35

Project Approval

An initiative that reaches Project Approval has already been tested and refined. Someone imagined it. Then teams shaped it, checked its business value, weighed it against other demands, and confirmed that people and budget exist to carry it out. Time and effort have already been poured into it and the organization has some level of commitment. However, Project Approval is not a rubber stamp. Executives are making a clear, public choice that this work will receive money, leadership attention, and real organizational focus. They are saying, without hedging, "We are doing this." This matters because once that decision is made, the project stops being an idea and starts becoming a promise. The result is an addition to the portfolio of projects that moves into execution with authority, defined backing, and no confusion about whether they should proceed.

The Purpose of the Executive Decision Meeting

An executive decision meeting has a single job: make the call. It's not a forum for endless debate, a place to redo analysis that should already be finished, or a workshop to redesign the solution on the fly. That work belongs earlier in the process. If that is happening then the project needs to be sent back to an earlier phase.

By the time a project reaches this room, the homework should be done. Leaders should have clear materials in front of them: a summary of value, defined scope, resource needs, architectural impact, risks, and how the work ties to strategy. They aren't there to invent the project from scratch, they're there to decide, based on solid preparation.

This distinction matters because when organizations blur the line between refinement and approval, meetings quickly become chaotic. The architecture team finds itself defending ideas that are still half-formed or decisions they thought had already been made. Business leaders feel caught off guard by details they have not seen before. Decisions get pushed "until next time", and the momentum is broken.

A well-run Project Approval meeting feels different. Materials are shared ahead of time. Participants know their role. The discussion centers on trade-offs and organizational fit, not on rewriting the proposal from the ground up. By the end of the session, the answer is clear: yes, no, or defer.

Who Must Be in the Room

Anyone sitting at the table needs real authority, the kind that unlocks budget, people, and time without running upstairs for permission. That usually means top leaders from the business units, finance, and IT. Enterprise Architecture shows up too, guiding the conversation and guarding alignment, but the final yes or no belongs to the executives on the hook for outcomes.

When the people with actual authority skip the meeting, everything slows to a crawl. You get half-answers and soft approvals like "subject to review by...", that kills forward progress. Why gather at all if no one can say yes? The rhythm only works when the right leaders are in the room, fully briefed, and ready to commit because momentum dies when ownership isn't clear. There's also a quieter cultural effect at play. When executives regularly show up and actively participate in Project Approval sessions, it signals that architecture and portfolio choices truly count. Over time, these sessions become woven into the way the company operate.

What Gets Reviewed

The content of the approval discussion should stay consistent from project to project. Predictability lowers friction and keeps conversations on track. In the Kettera approach, each initiative is presented through a structured summary that highlights several key elements.

First, strategic alignment. Which organizational goals does this project support? Is it revenue growth, operational efficiency, regulatory compliance, or improved customer experience? If the link to strategy is weak or indirect, that potential mis-alignment should be clear.

Second, value and cost.

What real payoff should we expect and when will it show up? Spell out the gains in concrete terms. Are we talking about higher revenue next quarter, lower support costs over two years, or faster delivery starting now? Then put numbers on the spending. How much cash goes out the door up front, and how much will it take to run and maintain over time? Leaders need to see the full money story, even if some of the figures are educated guesses.

Third, architectural impact.

Does this effort move us toward the future design shown in our current and planned system maps, or does it pull us off course? Will it clean up old, messy systems or add more complexity? Does it bring in a brand-new platform, double down on tools we already use, or create awkward workarounds no one asked for?

This is where Enterprise Architecture steps in. The job isn't to debate technical details for their own sake; it's to explain what those choices mean for the business in terms of risk, flexibility, speed, and long-term cost.

Fourth, risk and dependency

Are there firm regulatory deadlines to meet? Commitments to vendors that can't slip? Other roadmap projects this work relies on? When these questions surface late, they usually point to gaps in earlier planning. Present the details in a clear, consistent format and the meeting runs far more smoothly. Instead of scrambling to uncover basic facts, the group can focus on the real trade-offs and decision making.

The Nature of Executive Discussion

Imagine a mid-sized financial services firm weighing three initiatives: a mobile app redesign, a core system upgrade, and a new analytics platform for marketing. All three have cleared value analysis and prioritization. Resource planning shows only two can be fully funded this cycle.

The discussion in the Project Approval meeting will not center on whether the mobile app should use microservices or a monolithic architecture. That was settled earlier. Instead, executives will weigh timing and trade-offs. Is customer experience more urgent than modernizing infrastructure? Can the core system safely wait another year? Does the analytics platform generate revenue that offsets its cost?

These are not just technical questions. They are business calls informed by architecture. The role of EA in this meeting is to make the impact of each choice clear. If the core upgrade is delayed, what risk is surfaced? If the analytics platform is approved, how does it fit the target data architecture?

The strongest conversations are the ones where people speak openly. Leaders push back on shaky assumptions, dig into the numbers behind projected benefits, probe whether the plan really lines up with company goals, and call out risks that may have been glossed over. They test the idea from every angle.

Architecture leaders need to walk into these moments ready, and they should be if the previous phases were completed. They should explain their thinking in simple terms, respond to hard questions without bristling, and stay steady when their work is challenged.

Decision Outcomes: Yes, No, or Not Now

Every project review must end with a clear decision. If there are no answers coming out of this meeting then we haven't met the goal.

An approval unlocks funding, names who is accountable, and commits the organization's time and resources. The project joins the formal portfolio and moves toward transition and delivery. Scope limits and measures of success are written down so no one has to guess later.

Saying no sometimes is also important. It keeps the organization from stretching itself too thin and makes clear that the idea doesn't justify the cost or fit the strategy right now. A rejected proposal isn't gone forever, as it can return in a later cycle if circumstances shift.

Then there is the "not now" decision. This is different from a rejection. The project may be strategically sound but limited by timing, budget, or dependencies. In these cases, it goes back to the backlog with a clear explanation. The difference between "no" and "not now" should be written down; otherwise, teams keep lobbying informally and discipline starts to slip.

Clarity here protects everyone. Sponsors know where they stand. Resource managers can plan with confidence and architecture teams can keep a coherent roadmap.

Governance Without Bureaucracy

Governance can easily be perceived as red tape. When Project Approval turns into a maze of steps and forms, leaders look for side doors. They bankroll "small" efforts off the books or quietly call them routine operations just to dodge review. Before long, the rhythm of oversight falls apart.

Keep the oversight proportional. A tiny tweak to an existing tool shouldn't face the same scrutiny as a multi-million-dollar overhaul. Depth of review should match the size and risk of the idea. The Kettera approach builds this in, setting clear thresholds so teams know exactly how much detail to provide and when.

Clarity will lower people's defenses. People will cooperate when they see the reason behind the questions, like projected value or potential risk. They're far less annoyed when the question makes sense. The goal isn't control for control's sake, it's better choices, made with the best possible information.

Consistency across initiatives will build trust. When similar projects face different standards, confidence will fade, and executives start to wonder if politics matter more than analysis. Keep a steady rhythm in how decisions are made and much of that doubt will fade.

Documentation and Accountability

Once approved, a project's key attributes should be recorded in an easy-to-access portfolio artifact. This includes scope, expected benefits, budget range, timeline, sponsoring executive, and architectural considerations. The Architecture Decision Record (introduced earlier in the book) can also capture the major decisions made during approval.

Documentation matters because memory fades. Six months into execution, teams often disagree about what was originally promised. Was the goal cost reduction or revenue growth? Was the scope limited to one region or across the entire enterprise? Clear documentation stops revisionist history.

When everyone knows exactly who is on the hook, results improve. The executive sponsor is responsible for the business result such as the revenue gained, the cost saved, the promise delivered. The delivery leader runs the day-to-day work and makes sure the plan actually turns into something real. Enterprise Architecture keeps watch over alignment with standards and strategy. Enterprise Architecture also steps in when things drift off course.

Practical Example: A Healthcare Provider

A regional healthcare provider is weighing whether to expand its telehealth services. The proposal began in the operations team, then went through value analysis before rising on the priority list as patient demand grew and competitors rolled out similar offerings. Budget planning showed a clear trade-off: paying for telehealth would mean delaying a planned facilities upgrade.

At the Project Approval meeting, executives dug into patient access data, shifts in reimbursement policy, and the technical impact of scaling virtual visits. Enterprise Architecture confirmed that the proposed platform fit the organization's future integration model and could build on the existing identity management system. It also flagged a new requirement; stronger data governance controls to safeguard patient privacy.

By the end of the discussion, the executive team chose to fund telehealth and push the facilities upgrade back by a year. They recorded the rationale, noting projected growth in appointment volume and the architectural controls needed to reduce risk. With the trade-offs laid out in writing, the path forward was clear. That is the essence of Project Approval: conscious commitment.

Preparing for Transition

Delivery begins the moment the green light appears. Money gets assigned. People get pulled onto the team and the nuts-and-bolts design work ramps up. Oversight also changes shape to focus on guiding the project and helping it land well.

A pause between approval and action drains urgency, and attention drifts elsewhere. Sponsors get busy with other priorities and early assumptions age poorly. Projects need to roll straight into execution, cleanly and quickly, without pointless waiting.

During this transition, Enterprise Architecture helps keep things steady. As delivery planning speeds up, it makes sure the approved scope, standards, and integration approach stay consistent. Small adjustments that slip in during mobilization can quietly erode earlier decisions if no one is watching.

Key Takeaways

Project Approval is the moment when strategy turns into funded action. It takes place in a structured executive meeting built for clear decisions, not prolonged analysis. The right decision-makers must be in the room, supported by consistent materials that summarize value, cost, risk, and architectural alignment.

Each initiative receives a clear outcome: approve, reject, or defer. Documentation records commitments and prevents confusion later. Governance is scaled to the size and impact of the initiative, maintaining discipline without slowing progress. When this stage works well, the organization avoids scattered investments and political workarounds to avoid architecture. It builds a portfolio shaped by deliberate choices.

CHAPTER 36

Transition

By now, many of the hard choices are behind you. Teams have debated ideas and cut out the weak ones, tested the strong ones, weighed the payoff, and prioritized them. Money has been set aside and scope is clear. The architects have mapped the terrain and drawn the blueprint. But the truth is that all that planning counts for nothing if the work doesn't get kicked off.

Transition is the shift from intention to action; the point where roadmaps turn into tasks. It's where strategy stops being theoretical and starts affecting real schedules, real people, and real budgets. Done well, it feels smooth but done poorly, it will cause frustration and doubt.

This chapter looks at the formal step that moves an approved initiative into delivery: the Transition meeting with the implementation team. In that session, ownership changes hands and the role of the Enterprsie Architect changes. Architecture stops acting as the driver but remains involved to help ensure long-term success and alignment of the initiative.

The Purpose of Transition

Transition is not a ceremonial handoff. It is an operational checkpoint that makes sure what was approved is what actually gets built.

In many organizations, the gap between architecture and delivery is where projects start to fall apart. Architects create diagrams, roadmaps, and decision records and delivery teams inherit those artifacts, often without context. Assumptions get reinterpreted and constraints are forgotten. Deadlines drive shortcuts and months later, the solution barely resembles the original intent.

The Transition meeting exists to prevent that. Its purpose is threefold. First, it ensures a shared understanding between Enterprise Architecture and the implementation team. Second, it confirms that the project structure aligns with architectural decisions. Third, it formally shifts accountability for execution while keeping architectural oversight in place.

When Transition Happens

Transition takes place right after Project Approval and just before development work begins. The timing is important. If we hold the meeting too soon, key details may still be unclear. If we wait until coding is underway, and any changes can become costly and disruptive. Ideally, it happens at the point when:

- Funding is confirmed
- Scope and objectives are approved
- Key architectural artifacts are complete
- Implementation leadership is assigned

At this point, there should be very little ambiguity. The "why" is clear. The "what" is defined and the "how" has clear boundaries. The Transition meeting makes sure those boundaries are understood and accepted by the team building the solution.

Who Participates

Transition isn't just a conversation for Enterprise Architecture. The people who must actually deliver the work need to be in the room. They're the ones on the hook for getting the implementation across the finish line. In most cases, the group includes the individuals responsible for carrying the work through to completion, the doers, the decision-makers, and the ones accountable when results fall short. Typically, participants include:

- The Enterprise Architect leading the initiative
- The Solution Architect, if that role is separate
- The Delivery Lead or Project Manager
- Engineering or development leads
- The product owner or business sponsor
- Key representatives from operations or support, when relevant

Keep the group small enough to make clear decisions, but

varied enough to represent technical, business, and operational viewpoints.

This is not a status update. It's a hands-on working session. People should feel free to ask hard questions, challenge assumptions, and bring disagreements into the open.

1. Reconfirm Objectives and Outcomes

The Enterprise Architect or project sponsor needs to state the goals clearly. The key is to discuss what changes for the business and the people who run it. This team needs to understand the objective of the project so that when operating independently, the smaller decisions can be weighed against the larger goal and reduce drift throughout implementation.

For example:

- Cut order processing time from five days down to two
- Merge three outdated systems into a single, modernized platform
- Increase customer retention by 8% using digital self-service tools
- Lower support costs by reducing manual data entry and rework

When delivery teams understand the outcome, not just the feature list, they make better trade-offs later and can work more independently.

2. Review Architectural Decisions

Next is a focused walkthrough of the key architectural decisions. Not every diagram needs discussion, only the elements that constrain or guide implementation.

This usually includes:

- High-level view of the target architecture
- Common integration approaches
- Who owns the data and how it moves
- Agreed technology standards
- Security and regulatory needs
- Expectations for performance, uptime, and growth

Architecture Decision Records (ADRs) matter a lot here. They capture the thinking behind each key choice and spell out which options were considered. When engineers can see the logic that led to a decision, they're far less likely to debate the decision further. Understanding that debate has already occurred helps reduce resistance.

Delivery teams should probe some boundaries. Are these guardrails set in stone, or can they shift over time? That's not pushback; it's smart engagement. Some constraints won't budge like regulatory requirements but others can flex if there's a solid case and the right review. The real question is which decisions are permanent, and which are open to change?

3. Clarify Boundaries and Scope

Scope drift usually starts in the first sprint. A clear Transition meeting can head it off before it spreads.

Be explicit as a group about:

- What is in scope
- What is out of scope
- Known dependencies
- Assumptions that must hold true

Capture these on a shared board so everyone can see them. When the boundaries are visible, they're harder to ignore and easier to defend later. For instance, if an upgrade project assumes that an external vendor will deliver an API by a set date, that dependency should be clearly listed. If the assumption fails, the architecture may need changes.

4. Define Governance During Implementation

Architecture doesn't step away once the plan is approved. It stays involved, shaping decisions as the work unfolds and making sure the build still matches the intent. Be explicit about how that involvement works.

- Are there architecture check-ins tied to major milestones?
- Which kinds of changes must come back for architectural review?
- Who has the authority to approve exceptions to standards?
- How are risks raised and escalated when they can't be resolved locally?

Clarity here prevents friction later and avoids the deviations that derail programs. Without agreed guardrails, who decides when a "small tweak" becomes a real risk?

Cadence is a critical component to managing deviation from intention. Some efforts only need light oversight, maybe a monthly session to review progress and flag concerns. Others, especially those that touch core platforms or shared systems, demand tighter, more frequent engagement to stay on track. The goal of these touch points is not control. It is alignment.

From Approval to Active Implementation

When the Transition meeting wraps up successfully, the nature of the work changes. The project shifts from concept to action.

This shift includes several operational outcomes:

- The project is officially added to the implementation portfolio
- Delivery teams begin detailed planning or sprint execution
- Architectural documents are baselined
- Governance checkpoints are scheduled
- Success metrics are logged as performance targets

A Practical Example

A mid-sized financial services firm set out to consolidate its customer data platforms. The effort moved through idea generation, value analysis, and prioritization before receiving funding approval. The goal was to replace three disconnected systems with one cloud-based solution.

When the project entered Transition, the Enterprise Architect walked the team through the target data model, proposed integration patterns, and the selected cloud provider. At that point, the development lead flagged a concern: the planned integration method would add latency to a high-volume transaction process.

That single discussion reshaped the implementation plan. Instead of applying the same pattern everywhere, the team introduced a hybrid integration model for that specific workflow. They documented the decision and scheduled governance checkpoints at each major integration milestone.

If the latency problem had appeared three months later, the rework would have been expensive and disruptive. Catching it during Transition made the adjustment manageable and timely.

Common Failure Patterns

Despite its simplicity, Transition is often mishandled. A common mistake is treating it like handing out documentation. Architects send slides. Developers flip through them. No real discussion takes place and weeks later, misunderstandings surface.

Too much detail will actually degrade the value of this meeting. When a meeting slips into the weeds of technical design, people stop thinking about the bigger goal. They argue over specs, diagrams and edge cases. The original purpose of getting everyone pointed in the same direction will fail. A transition meeting should lock in shared intent, not turn into a line-by-line design critique.

Missing decision-makers creates a different kind of problem. When the delivery leads in the room can't actually say "yes" or "no," the discussion sounds productive but goes nowhere. Ideas are kicked around but nothing sticks. Put people at the table who have the power to commit and who will stand behind the call once it's made.

These patterns are avoidable with discipline and preparation.

The Human Side of Transition

In addition to the mechanics of the process, Transition is about trust. Delivery teams need confidence that architecture is practical and useful, not theoretical. Architects need confidence that the build will respect the guardrails. This meeting makes that relationship possible.

When architects sit down and explain their thinking like why they chose one platform over another or why a specific data standard actually matters, pushback tends to decrease. Developers like seeing how their work fits into the larger system. Sponsors relax when they see tech leads and business leaders moving in the same direction. Over time, people stop viewing architecture as a gatekeeper and start seeing it as a guide. That perception spreads across teams, shapes decisions, and influences culture in ways that last.

Transition Within the Enterprise Architecture Cadence

In the broader Kettera EA Methodology, Transition is the point where work moves from strategic cadence into delivery cadence. It signals that planning gives way to execution and that direction turns into action.

Earlier in this section, we walked through how ideas advance through a structured path we call the Innovation Pipeline: Idea Generation, Refinement, Value Analysis, Prioritization, Resource Planning, and Project Approval. Each stage applies a different lens to test assumptions and clarify value. By the time an initiative reaches approval, it has been examined from multiple angles, so the organization is far less likely to commit funding and time to the wrong effort. Transition keeps the organization from building the right initiative the wrong way.

Key Takeaways

Transition is the formal step that moves an approved initiative into active delivery. In a well-run Transition meeting, teams:

- Reconfirm the business objectives
- Revisit the key architectural decisions
- Clarify scope boundaries and dependencies
- Set governance expectations for delivery
- Align implementation teams with enterprise standards

The result is that we have active execution based on agreed

architecture and measurable results. Without a meaningful Transition, even well-designed strategies can fall apart during execution. With it, Enterprise Architecture stays connected to the delivery process.

Conclusion

CHAPTER 37

Key Take-Aways

We've covered methodology, artifacts, governance, and execution. Now stop and consider one basic question: what actually matters? Enterprise Architecture can feel huge. It stretches across strategy, day-to-day operations, data, applications, and the underlying technology stack. That sounds like a lot because it is. But the ideas at the center are straightforward. They work in the real world and they are repeatable. They reflect how companies actually make choices.

Enterprise Architecture, as explained through the Kettera EA Methodology, isn't academic theory. It's a practical system you can run. In a nutshell, EA exists to create clarity, alignment, focus, and steady, managed change. Every artifact, every roadmap, every standards document and every governance meeting is a tool designed to achieve this.

If those four things are happening, the architecture is doing its job. If they aren'tthen we need to refine and customize the methodology.

Here are the key ideas I hope stay with you.

Enterprise Architecture Is About Organizational Success

The first and most important takeaway is this: Enterprise Architecture exists to support organizational goals, not technology goals. It begins with strategy. It starts with understanding what the business is trying to achieve, how it creates value, and where it plans to compete.

That is why this book looked at marketing domains, operations domains, and finance domains before going deeper into applications and platforms. Technology is an enabler. It can speed up growth, lower costs, improve customer experience, or create entirely new offerings but only when it is deliberately aligned with the organization's strategy. Without that alignment, even well-designed systems can pull the enterprise in different directions.

A company pushing into a new region often pours money into brand new infrastructure but that's not always the real gap. The bigger need may be systems that let customers sign up at scale, support teams that speak the local language, reports that meet local rules, and dashboards that show how the region is actually performing. What's the point of faster servers if customers can't even onboard smoothly? Enterprise Architecture puts technology choices inside the bigger business goal. It forces one simple question first: What are we trying to get done? Only then does it ask what to build.

When Enterprise Architecture turns into an inward-looking IT project cut off from business goals, it stops mattering. The whole reason it exists is to connect technology efforts to business outcomes.

Methodology Matters More Than Framework

Earlier in the book, we separated frameworks from methodology. Frameworks offer structure: taxonomies, viewpoints, and reference models. They help organize thinking, but they do not show you what to do next. The Kettera Architecture Methodology is about movement. It sets a steady cadence for generating ideas, refining them, analyzing value, setting priorities, planning resources, securing approval, and guiding transition. That focus on motion is what drives results.

A methodology tackles the practical questions frameworks often leave unanswered. When should architecture step in? How are ideas judged? Who actually makes decisions? What happens to roadmaps when conditions shift? Without clear answers, Enterprise Architecture turns into ceremony with review boards approving slide decks while real decisions happen elsewhere.

When the methodology is active, architecture shows up where decisions happen. Architects work within the cycle of change instead of reacting after the fact.

Constraints Are Not the Enemy

Constraints, whether financial, regulatory, or technological, are not barriers to progress. They are design inputs. They set the boundaries within which smart decisions are made. Architects who push back against constraints often produce elegant models that can't be implemented. Architects who accept them, produce solutions that work. A limited budget forces prioritization. A compliance requirement clarifies how data is handled. A legacy platform can dictate integration choices. These realities shape architecture in practical ways.

Imagine designing a future-state architecture without recognizing contractual vendor commitments or regulatory reporting obligations. It would look impressive on paper and fail in practice. By contrast, acknowledging constraints early allows architects to propose phased transitions, realistic roadmaps, and focused modernization efforts.

Enterprise Architecture isn't about sketching some flawless future. It's about helping an organization get better step by step, working with the limits it actually has such as tight budgets, legacy systems, office politics, and many others. This work is practical. It means steering change in a messy, real environment, making smart trade-offs, and choosing improvements that fit the world as it is, not as we wish it was.

Artifacts Are Tools, Not Deliverables

Capability maps, current and future state diagrams, roadmaps, application portfolios, data relationship diagrams, standards, decision records. This book has explored a wide range of artifacts. They all serve a purpose, but none of them stand on their own.

Their role is to help people make decisions. A capability map shows what the organization needs to be able to do. A roadmap lays out change across the next 12 to 24 months. An application portfolio highlights duplication and risk. An Architecture Decision Record captures the reasoning behind a choice so future teams aren't left guessing.

Problems arise when the artifact becomes the objective instead of the aid. Thick binders and intricate models can give the comforting sense that real progress is happening but if they don't shape funding choices, steer solution design, or influence priorities, they're just decoration. Effective architects adjust artifacts to fit the audience. Executives look for clear trade-offs and investment impact. Delivery teams want standards and integration guidance they can apply right away. Governance groups need a visible thread from strategy through to implementation. Format matters far less than whether the artifact actually helps someone act.

When in doubt, ask: "What decision will this artifact help someone make?" If you don't have a good answer, rethink its purpose or skip it entirely.

Architecture Is a Continuous Conversation

Enterprise Architecture isn't a one-time effort with a kickoff and a finish line. It never stops. It's a steady, living discussion about how the organization changes and why those changes matter. When that conversation stops, confusion creeps in, priorities drift, and teams pull in different directions. Rhythm matters because the work never truly pauses, it just moves to the next decision.

The innovation pipeline makes this clear. Ideas surface from all over the business. The strongest ones get shaped and tested. Leaders examine the value, weigh trade-offs, and choose what deserves attention. Money and people get assigned. Work begins, results appear, and adjustments follow. Then it all loops back around. Architecture stays involved the entire time, guiding choices, and keeping the bigger picture intact. It provides structure without crushing creativity. The goal isn't control for its own sake. The goal is steady alignment as the business keeps evolving.

This conversation needs structure with clear roles, agreed upon checkpoints and written standards but it also needs room to adapt. Markets shift, leadership turns over and technologies evolve. If the architecture function is too rigid to respond, it won't stay relevant for long. The idea of a "concrete structure with flexibility" captures that tension. Governance should be defined without sliding into red tape. Standards should guide decisions without turning into dogma when circumstances change. Architecture exists to keep the organization moving, not to cause rigidity.

Roadmaps Create Confidence

One of the most practical tools in this book is the 12- to 24-month roadmap. Long-term visions are inspiring, but near-term sequencing builds confidence. Leaders are more willing to fund initiatives when they can see how projects connect and how value builds over time. A roadmap turns strategy into action. It shows dependencies between systems. It highlights capability gaps. It makes trade-offs clear. When executives see that modernizing the data platform enables better analytics, which in turn supports targeted marketing campaigns, investment decisions become clearer.

Equally important, roadmaps reduce chaos. Without one, projects compete for attention without shared context. With a roadmap, initiatives align around a clear direction. Adjustments still happen, but within an agreed framework. The roadmap isn't a guarantee that everything will stay the same. It's a mutual agreement about where we're trying to go and how we expect to move forward.

Plans evolve and priorities shift. New information shows up when you least expect it. A roadmap simply lays out the direction we believe makes sense right now. Think of it as a working guide, not a contract. It reflects today's best thinking and a common view of progress.

Decisions Must Be Documented

Architecture Decision Records (ADRs) look straightforward, but they address a stubborn issue in most organizations: shared memory fades. Teams change. Leaders cycle through roles. Vendors arrive and leave. When no one writes decisions down, the logic behind major architectural calls vanishes.

Put the decision such as options considered, path selected, and the reasoning on record, and future discussions will start from those facts, not vague memories. Teams can test old assumptions instead of reopening arguments no one fully remembers. Written records also create accountability by showing that choices were deliberate. Over time, a consistent archive of ADRs turns into a real asset. It shows how the company weighs trade-offs, handles risk, and sets standards.

Enterprise Architecture Is a Leadership Function

Although Enterprise Architecture often sits within IT, its influence reaches beyond technical boundaries. It connects strategy to execution. That connection is a leadership responsibility. Effective architects communicate clearly with executives, product owners, engineers, finance teams, and operations managers. They translate strategic goals into capability requirements. They translate technical constraints into what they mean for the business. They ask difficult questions about value, cost, and risk.

Leadership in this context does not require authority over budgets or headcount. It requires credibility. Architects build credibility by understanding the business, delivering practical guidance, and being consistent in their recommendations. When Enterprise Architecture earns trust, it becomes a partner in shaping the organization's direction instead of an after-the-fact reviewer.

The Kettera Perspective: Structured Motion

Discipline in motion defines the Kettera EA Methodology. Architecture doesn't create a static state; it's constantly moving. It moves alongside the business instead of trailing behind it. Every step of the cadence weaves architecture straight into how choices are made. Bright ideas don't advance just because people feel excited about them. They're shaped, tested for real value, ranked against other demands, and mapped to the time and people actually available. Architecture ties all of that together so it makes sense.

Perfection isn't the goal. No methodology wipes out doubt or disagreement. What this one does is cut down the chaos. It makes decision paths visible. It links working documents, oversight, and delivery into one connected system with fewer loose ends and surprises.

Key Take-Aways

As you think back on the ideas in this book, keep these principles in mind:

- Enterprise Architecture serves business strategy first and technology second.
- Methodology determines effectiveness more than adherence to any single framework.
- Constraints define intelligent design; they do not prevent it.
- Artifacts must inform real decisions or they have no purpose.
- Architecture is continuous, embedded in the organization's cadence of change.
- Roadmaps translate strategy into actionable sequencing.
- Documented decisions build institutional memory and accountability.

- Architecture is a leadership discipline grounded in credibility and clarity.

At its heart, Enterprise Architecture is about making change deliberate. It aligns investments, systems, and processes so they move in the same direction instead of drifting apart. Done well, it cuts through fragmentation, informs priorities, and strengthens the link between vision and execution.

Organizations aren't getting simpler. Growth, competition, regulation, and new technology keep adding moving parts. What they need is structure without rigidity, governance without paralysis, and clear thinking applied steadily over time.

In the final chapter, we look ahead. What does it mean to begin or renew your Enterprise Architecture journey? How can you put these principles into practice in your own organization?

CHAPTER 38

What's Next

You've almost finished the book but the work itself is just getting started. Enterprise Architecture isn't something you study once and check off a list; it's a craft you revisit again and again. The Kettera EA Methodology gives you structure and clarity, but those benefits only show up when you use it inside an actual company, with budgets, stakeholders, and shifting priorities. So what happens now?

Start Small, But Start Intentionally

Many organizations try to "roll out EA" as a sweeping initiative. They introduce a framework, book a few meetings, and send around templates. A few months pass, and little actually changes. That outcome isn't surprising. When architecture isn't tied to a real business problem, it feels theoretical and theoretical work rarely holds people's attention.

A better approach is to begin with something that already matters. Look for an initiative that is underway: a digital transformation effort, an ERP replacement, a customer experience redesign, or a push to cut costs. Apply the Kettera methodology there. Make the goals explicit. Capture how things work today. Describe a practical future state. Then lay out a 12- to 24-month roadmap that fits existing budgets and staffing limits. When architecture is connected to work people can see and measure, it becomes part of the conversation rather than background noise.

By starting this way, you do two things. First, you create value quickly. Second, you show that architecture isn't red tape. It's structured thinking applied to real decisions. That credibility becomes the base for expanding EA across the enterprise.

Adapt the Methodology to Fit Your Organization

Kettera lays out a rhythm, key deliverables, and clear decision paths, but your company runs on its own culture, power structures, and tolerance for change. Some businesses keep decisions at the center; others push authority out to the edges. Some demand robust documentation for every initiative; others would rather hash things out in a room with a whiteboard.

In fast-moving cultures that hate paperwork, stick to the essentials that actually create clarity: clear capability maps and a roadmap. Keep Architecture Decision Records short and useful. In more regulated environments, tighten the reins. Add stronger governance checkpoints, formal review boards, and clear links between architecture standards and audit or policy needs.

Don't lose the purpose while tweaking the packaging. The cycle still matters: generate ideas, refine them, test for value, set priorities, line up funding and people, approve, then hand off to delivery. That steady push from concept to value is what turns architecture into a living discipline instead of a bunch of unused documents.

Go Deeper in a Domain

Some readers will naturally gravitate toward one area over others. A business leader might choose to sharpen their understanding of business architecture, while a technologist may be more interested in application or data architecture. Focusing on one or more of these areas will provide an architect with important depth, as Enterprise Architects are usually expected to to be knowledgable in one or more of these areas.

If business architecture speaks to you, revisit the early chapters on organizational goals and domains. Rework your capability maps and test whether they truly reflect how the company creates value. Make it explicit how marketing, operations, finance, and IT support the broader strategy. Then connect initiatives directly to the capabilities they are meant to strengthen. Organizations often stumble not because their technology falls short, but because they lack clarity about which capabilities actually set them apart. When alignment improves, investment decisions tend to change in noticeable ways.

If application architecture captures your attention, take a close look at your application portfolio. Which systems overlap? Which are barely used? Where does integration complexity add risk or drive up cost? Rationalizing even part of an application portfolio can free up budget and make future initiatives easier. Pair this with clear standards and reference frameworks so new projects don't repeat old mistakes.

Data architecture offers another rich area of opportunity. Clean data relationships, clear ownership, and consistent definitions of key entities like customer, product and order, can dramatically improve reporting and analytics. When data models align to business capabilities instead of departmental silos, decision-making gets better. Few investments pay off more over time than disciplined data architecture.

The goal is not to specialize at the expense of the whole. Enterprise Architecture is still enterprise-wide. But deep expertise in one domain strengthens your ability to shape broader decisions.

Strengthen the Cadence

An architecture cadence is simply a steady path from a rough idea to real-world rollout. If you want a concrete next move, make that rhythm part of how your organization operates every day. Set regular checkpoints. Decide exactly who helps shape and challenge new ideas. Define how you judge value and how you log priority calls. Add clear timelines for each step so it's clear to all stakeholders. People need to see how it works and when to be involved. Walk project managers, product leads, and business sponsors through the flow, showing them where their decisions fit. With repetition, the cadence eventually becomes part of the company's muscle memory.

Develop the People Around You

Enterprise Architecture is often seen as the job of a small, centralized team, but that setup rarely scales well. A healthier long-term model spreads architectural thinking across many roles instead of concentrating it in one group. Help solution architects tie their designs directly to enterprise standards and roadmaps. Ask business analysts to trace requirements back to business capabilities. Partner with finance teams so architectural implications show up in funding conversations. Product managers also gain perspective when they understand the gap between today's state and the intended future state.

Consider forming informal communities of practice to keep the conversation going. A monthly architecture forum can give teams space to share recent decisions and what they learned along the way. Review Architecture Decision Records together and talk candidly about trade-offs. Over time, these habits build a common language and a sense that architecture is everyone's responsibility, not just a central team's concern. Additionally, as architectural literacy spreads, resistance tends to drop. People stop seeing EA as an outside control system and start viewing it as a shared way to make better decisions.

Measure What Matters

Asking "What next?" isn't enough. You also need to ask, "How will we tell if this is actually getting better?" Architecture maturity has nothing to do with how many diagrams get created. Real progress shows up in business results. Look for fewer duplicate applications across the business. Check how long it takes to move from a raw idea to an approved initiative. Compare funded projects to the goals leaders say matter most. Watch technical debt over time and see whether integrations are getting simpler or more tangled. Even soft signals count, like whether executives trust the technology roadmap. This data should change what you do next. If prioritization meetings keep ending with no decisions, improve how you score and compare value. If nobody looks at the roadmap, make it easier to find or clean it up so it's easier to consume.

Expand Your Influence Thoughtfully

When architecture proves its worth in one domain or initiative, new doors tend to open. Growth is tempting but scaling too fast can spread out your attention and erode credibility. Expanding without the capacity to engage deeply often results in surface-level involvement and weaker outcomes. Be deliberate about what you take on next. Finance might ask for help aligning capital planning with technology investments, or operations may want guidance in simplifying process automation tools. Don't change your method just because the audience has changed. Start by clarifying the current state, define a practical future state, and map a path forward tied to measurable results. Over time, architecture becomes less about stand-alone diagrams and more about shaping how the organization thinks about change.

Look Beyond the Organization

Your path doesn't end with improving things inside your own company. Talk with architects at other firms and see how they handle governance and decision rights. Look closely at how different industries organize their architecture practices and map capabilities. Lay those maps side by side and study them. Many patterns will repeat, but the differences can be revealing.

Professional associations, conferences, industry forums, and even active online communities expose you to fresh thinking you won't find in-house. Academic studies add context about strategy and organization design and you can use them to test your assumptions. Mentoring junior architects can also be a very valuable endeavor, not just for the mentee but for you too. Explaining your reasoning reinforces it and causes you to rethink your assumptions.

The Future of Enterprise Architecture

Technology will keep evolving. Artificial intelligence, automation, and new platform models will reshape how organizations operate. Yet the core questions stay the same: What are we trying to achieve? What capabilities do we need to strengthen? How do we move from where we are today to something better?

Enterprise Architecture is here to answer those questions in a structured way. The Kettera methodology offers a practical framework, rooted in business goals, supported by clear artifacts, and strengthened by a repeatable rhythm. It is not overly theoretical or narrowly technical. It helps organizations think clearly and act with purpose.

Your next step is simple: practice. Choose one initiative and apply the method. Share what works and adjust what doesn't. Over time, the discipline becomes part of the organization's culture. Carry the most important ideas forward: stay tied to company goals, spell out where you are and where you're headed, choose priorities with care, and keep a steady rhythm in how you implement your practice. Do that, and you'll be ready to drive change that actually matters.

About the Author

Jeff Davis is the Founder and CEO of Kettera Solutions, LLC, bringing nearly 3 decades of experience across enterprise architecture, software development, innovation management, and more recently, AI enablement. He is the creator of the Kettera Architecture Methodology, a structured, repeatable approach designed to bring clarity, consistency, and adaptability to enterprise architecture.

Recognizing a gap between theory and real-world practice, Jeff wrote this book to help current and aspiring enterprise architects navigate complexity with confidence. His methodology emphasizes a balance between rigor and agility, enabling teams to move quickly while maintaining architectural integrity.

Based in Dallas, Texas, Jeff continues to work with organizations to modernize their architecture practices and effectively adopt emerging technologies.

www.ingramcontent.com/pod-product-compliance
Lightning Source LLC
LaVergne TN
LVHW052023200926
848481LV00005B/1570
9798995864301